IEE MANAGEMENT OF TECHNOLOGY SERIES 19

Series Editors G. A. Montgomerie
                B. C. Twiss

CONTINUING PROFESSIONAL DEVELOPMENT

a practical approach

Managing your CPD as a
professional engineer

Other volumes in this series:

CONTINUING PROFESSIONAL DEVELOPMENT

a practical approach

Managing your CPD as a professional engineer

John Lorriman

The Institution of Electrical Engineers

Published by: The Institution of Electrical Engineers, London,
United Kingdom

The Institution of Electrical Engineers,
Michael Faraday House,
Six Hills Way, Stevenage,
Herts. SG1 2AY, United Kingdom

British Library Cataloguing in Publication Data

A CIP catalogue record for this book
is available from the British Library

ISBN 0 85296 903 1

Printed in England by Redwood Books, Trowbridge

Contents

List of figures

Foreword

This book came about as a result of many years of personal commitment to continuing professional development. The reason I feel so strongly about the topic is because I have observed so many engineers in so many organisations being misutilised and underutilised.

Engineering can be the most exciting career in the world, and in such an environment CPD should be no problem at all. Sadly, all too often it is not as prevalent as I would like it to be. So what has gone wrong?

For a start, individual engineers tend to look to others to structure both their career and learning. A key message in this book is that each one of us needs to take personal responsibility for our own competence and career development—and to a very high degree, as well as continually.

Another message is that organisations need to redefine the role of managers in a fundamental way if CPD is to be successful. It needs to be accepted that the prime role of a manager is to coach and develop his or her staff. And that is, frankly, a fundamental challenge in most organisations.

Thirdly, the providers of CPD need to rethink their approach, so that the learners are best able to relate their learning to specific competence and career development objectives.

Fourthly, the professional institutions need to do as much as possible to learn from each other's successes and good ideas. They face many exciting challenges, such as how to make best use of the Internet, groupware and software for personal development plans.

This book is aimed at readers from all four groups—individuals, organisations, providers and professional institutions. CPD needs to involve all four continually, each interlinked and mutually supportive, and the contribution from each group is covered as far as possible in each chapter. Therefore, from whichever perspective you read this book, please bear in mind the contribution you can either receive from, or even better give to, the other three groups. At the end of each chapter, think about the learning you may have had from that chapter in as broad a context as possible.

Finally, CPD has to be driven very much at an individual level. So how about asking yourself at the end of the day the following three questions:

- what went well?
- how could it have gone better?
- what did I learn?

Good luck—and I hope you enjoy reading this book. In any event, please send your comments, as well as your own CPD experiences, to me either care of the IEE's professional development manager or on the Internet at Jlorriman@iee.org.uk.

John Lorriman

Chapter 1
What is continuing professional development?

'You own the world's most powerful computer—your brain'

I am starting to write this book in an aircraft, 37 550 feet above China, on my way to spend two weeks in Hong Kong doing training accreditation visits for the Institution of Electrical Engineers (IEE). Is this part of my continuing professional development (CPD)? Not half!

CPD can have many different dimensions—limited only by your imagination—ranging from the sort of experience described above, to reading a book or an article. What matters is that:

- you use a variety of methods for your CPD;
- these methods are complementary and give you good all round personal and technical development;
- whatever approaches you use, they actually work for you;
- you find that your CPD is fun!

This book is aimed at helping you to structure, target and develop your CPD much more effectively. But at the start we need to be clear about what CPD actually is!

1.1 Ask yourself what you mean by CPD

As I sit on this aircraft, I find that the person sitting next to me is an electronics engineer, working for a very successful American software company in Ireland. Previously she worked for a major electronics company based in Holland, and I decide to ask her what she considers CPD to be about. Here are some of her thoughts:

- it is about letting people do the work—it is about trust;
- it is about changing jobs and challenging yourself;
- you need to learn from your mistakes as well as your successes;

- there is a need to maximise the ability of organisations to change and to keep up with the Japanese;
- CPD must be about maximising the learning process.

Why not take a sheet of paper and, using the format below, write down your own ideas about what CPD is and what it might be about in the future? Alternatively, how about creating a software document with this format on your PC and then updating it as new thoughts occur?

What CPD means to me	What my CPD might be about in the future

1.2 Defining CPD

It may be helpful to see whether one or two definitions help to clarify your thoughts about what CPD actually means to you.

A definition of CPD which has generally been accepted by The Engineering Council's member institutions is:

'The systematic maintenance, improvement and broadening of knowledge and skill, and the development of personal qualities necessary for the execution of professional and technical duties throughout the practitioner's working life.'

On the other hand, you may like a much simpler definition, which I rather prefer myself: 'Steal ideas shamelessly.'

This definition, at least, has the advantages that it is short and memorable, as well as concentrating the mind on the need to maximise the ways in which we obtain new ideas from others. One recent redefinition of this is: 'Steal ideas with pride'!

1.3 How much of your potential are you using now?

Ask yourself this question: 'How much of my inherent potential have I used up to now?' Is the answer in your case high or low? For many people it is as low as twenty per cent.

What, then, might be the level of your potential release? How many ideas have you had during your life which have not come to fruition? Which career development aims have you had that are still unrealised? What skills do you have that you really want to develop further? Which new skills do you want to develop?

Take a clean sheet of paper and write down your ideas. Ask yourself whether you are really satisfied with your career so far. How could you raise your future potential? What are the real barriers which stop you from doing this? How many of these barriers are really in your mind?

Your brain is almost certainly the most complex structure on earth— and maybe in the universe. It has some 30 billion neurons, or nerve cells, which link together in thousands of different ways—so have a high regard for the almost unlimited potential which this gives you. On the other hand, respect its need for rest; the term powernapping is becoming fashionable these days. Powernapping is the habit which many people, including Richard Branson of Virgin, many IBM managers and senior executives across America and Asia-Pacific countries, are adopting of taking a few moments during the working day to rest the brain by switching off completely. Professor Jim Horne, director of the Sleep Research Laboratory at Loughborough University, says: 'We have a natural disposition to sleep twice a day. All over the world people are programmed to sleep twice: a short sleep in the middle of the day and a major sleep at night.' I have adopted this approach for very many years and find that it is an excellent way of not only unwinding and refreshing my brain, particularly at the end of a working day, but also bringing out many new ideas of which I had only been half aware previously.

1.4 How good have your learning processes been?

There are many different ways in which you can develop your CPD— sometimes from learning from your successes and sometimes from learning from your mistakes.

Use Figure 1.1 to draw a chart showing how your learning has developed your competence over the course of your career, starting with the early years immediately after you obtained your initial qualification. The vertical axis shows the progression of your career over time, with the start at the bottom. To the right draw a line showing how your cumulative competencies have developed as a result of positive experiences and successes, while to the left of the vertical axis draw a line showing how your learning and competencies have developed from the mistakes that you have made and any negative experiences which you might have had.

With hindsight, could you have developed your skills and knowledge even faster? In my own case, I realised two years after I had graduated

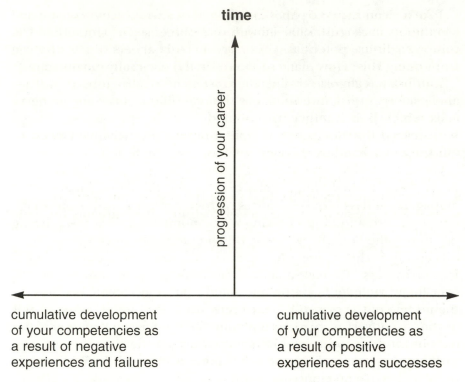

Figure 1.1   *Cumulative development of your competencies*

that I had basically had one year's experience repeated twice, rather than two years of useful personal and professional development.

Have you learnt more from your successes or your failures? If the amount of learning from your failures has been small, then have you taken enough risks in your career or have you been over cautious?

Give some time now to thinking through these issues. Ideally, sit down for an hour or two with a colleague, your mentor (if you have one), or a friend and discuss your thoughts on these issues.

1.5 How much more of your potential could be released in future?

Professor Charles Handy has suggested that, by the time we die, most of us have released no more than twenty to twenty five per cent of our inherent potential. What would we, our organisations, our countries and even the world look like if that could be raised to thirty five, fifty or even ninety per cent, he asks?

A survey conducted by MORI on behalf of management consultants Proudfoots in August 1996 showed that only sixteen per cent of UK company directors feel that they are making the best use of their employees. This figure suggests that even the most senior managers in UK industry recognise that much more needs to be done to release employee potential, but ultimately the real responsibility for doing so must rest with each individual.

In order to release more of your inherent potential, you need to think about a number of issues, such as:

• how much of your real potential have you realised so far?
• what would you really like to do with the rest of your life?
• what barriers stand in the way of realising these dreams for your future?
• how can you begin to take steps to maximise the likelihood that these dreams will become reality?

Too many people fail to understand that the major barriers to achieving what they really want lie in their own mind. Lack of self confidence or lack of determination are much more usually the reasons for under achieving than any failure on the part of managers, employers and so on. On the other hand, as we will see in the next chapter, both managers and employers have a key role to play in helping individuals to realise their ambitions, but the starting point has to be in the minds and imaginations of each one of those individuals.

1.6 Committing yourself to CPD

Think about the implications of the three words of CPD.

Continuing means that you need to commit yourself to a lifelong learning process. Achieving your initial professional qualification is just the start of the process of self development.

Here in Hong Kong I have now finished the first week of the Institution of Electrical Engineers' training accreditation visits. On one of them, we were returning on a company ferry from an accreditation visit to Hong Kong Electric, based on Lamma Island just off Hong Kong. Our guide, a young graduate who is now a safety and training engineer, explained that he was going on to an evening class at Hong Kong Polytechnic University as part of a three-year evening MSc that he is doing at his own expense three evenings a week. Furthermore, he remarked that perhaps thirty per cent of the 200 or so people travelling with us on the company ferry back to Hong Kong were going on to evening courses that night, almost all at their own expense.

Over the weekend we took a ferry to Lantau Island, where the new airport is being built. At the ferry terminal there was a rack filled with leaflets about self improvement, and education and training courses, which young people were queuing up to pick up and read.

How continuing, therefore, is your commitment to your self development? Can you and your organisation compete with the energy and commitment shown here in Hong Kong?

Professional implies that you will take an approach to your self development which will merit the respect of your fellow professionals. You need to analyse, perhaps, just how professionally you have developed your competencies and your career so far. Have you, for example, participated actively in your institution? Have you taken the trouble to read your institution's publications and found as many other ways as possible of keeping up to date?

The European Union has estimated that two-thirds of the technologies which will be invented by the year 2000 had not been invented in 1990, and yet three out of four people working at the beginning of the 1990s would still be active by the end of the decade. In many areas it has been estimated that twenty per cent of an engineer's knowledge becomes obsolete each year.

How fast is your knowledge becoming outdated? What can you do to keep it up to date? And what could you do to pass on your expertise to help others to keep updated?

The word *development* should cause you to ponder on the shape of the graph you drew in Figure 1.1. How effective has your development been so far? How could it be made better—much better—in the future? What help do you need? What do you need to do? How can others help you?

1.7 Maximising the learning process

By now you should have asked yourself a lot of questions. In some cases you will already have answers with which you are comfortable. In others you will need more time to think through the issues which have been raised.

It is likely, however, that you have made some notes on your thoughts, and possibly on your discussions with others. Keep these filed safely because they are the start, if you do not already have one, of a personal development plan. If you already have such a document, then use it to collate and collect your conclusions and commitment to future action. If you do not have a personal development plan yet, then we will look at different ways of producing one in Chapter 5.

The key question, however, to finish with in this chapter is this:'How can I maximise my learning process?'

If you are learning as fast as possible, then you will be becoming more competent that much faster. If you are at the forefront of your particular area of expertise, then one way or another you will be rewarded, whether by a salary increase, by job satisfaction, by better career progression or by enhanced respect from your colleagues. And that is what continuing professional development is really all about.

Chapter 2

Getting it all into context—
managing your CPD

*'Whether you think you can—or whether you think you can't —
you are probably right!'—Henry Ford*

There are three key elements in successful CPD—whether from an individual or an organisational point of view.

First, there must be a *missionary* zeal by each and every individual for their self development. This must be based on a clear understanding of their key competencies, and all their learning must be directly linked to maximising the speed with which they develop these competencies. They need to use personal development plans (PDPs) and to have a clear idea of how they will develop their future career.

Secondly, organisations must redefine the role of their managers, so that coaching and developing staff is the most important activity. This must be the basis on which managers are appointed, trained, developed, appraised, rewarded and promoted.

Thirdly, the organisation must be a learning organisation. This means two important things: that the organisation maximises its learning processes and that its value system is cohesive. For example, will managers be committed to coaching and developing their staff if they are not predominantly rewarded and promoted on the basis of their ability and success in doing this?

2.1 Lorriman's Windows

In my view, each of the three factors above is a window of opportunity—which with complete immodesty I have called Lorriman's Windows. These are shown diagrammatically in Figure 2.1.

My argument is that if any one of these three factors is missing, then an organisation will have difficulty in becoming, or staying, world class. And if your organisation does not aim to be world class, then it should

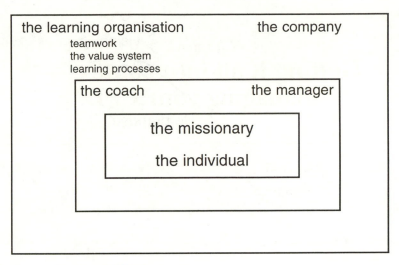

Figure 2.1   Lorriman's Windows

question why it is in business at all, because sooner or later another competitor will put all three windows in place.

Managing your CPD, therefore, implies that you choose to work for an organisation which does indeed recognise the importance of each and every one of these three windows. But the most important starting point of all is that you recognise your own responsibility in all this.

2.2 Managing your own CPD

Far too many individuals look to others for the development of their CPD, whether in terms of their competencies or their careers.

For example, the larger the employer, the more there seems to be a tendency to look to the training department or personnel department to provide CPD. But this culture of dependency is very damaging, because it shifts the responsibility from the individual to the organisation. And yet no approach to CPD can work unless each and every individual in an organisation is totally committed to their own self development.

At the very least this requires a commitment to structuring and recording your own learning activities, and some degree of targeting of your career development. We will look in Chapter 5 at how you can develop and use your own personal development plan in order to do this.

In Chapter 1 you started asking yourself some important questions. One of the most important, perhaps, is to ask yourself about the

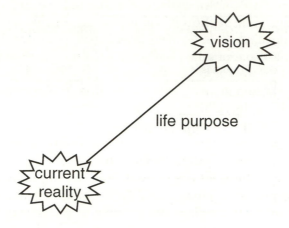

Figure 2.2   Current reality/vision model

visions you have—those unfulfilled, or half-fulfilled, dreams. Figure 2.2 illustrates the link between the current reality of where you are now, and the vision you have for your future.

For many, if not most, of us, our visions are far too limited, and we are almost all able to realise far greater things. By talking your ambitions through with a mentor, coach or friend you are likely to become more aware of these ambitions. By verbalising them, you will feel more confident and committed to achieving them.

These ambitions can be short term. In one instance a senior manager decided that his great ambition was to achieve an Advanced Driver's licence. Until then he had assumed that he was far too busy to find the time to study for this qualification but, after using his personal development plan to analyse that this was a key unrealised ambition, he talked things over with his secretary. They agreed that he would leave the office each Thursday afternoon at 4.30 p.m. precisely in order that he could attend the advanced driver course. He was so committed to this that he even decided to miss the annual Christmas cocktail party given by his company chairman, which normally all senior managers would attend as part of their status in the company. As a result of being able to achieve this long-held dream, this senior manager felt able to give even more energy to the company than he had before. This is a good illustration of how finding a better balance to your life can also provide good returns to your organisation.

When Dr. W. Edwards Deming, the great American quality expert, first went to Japan in the early 1950s, he drew a picture on the whitewashed wall of the room where he ran his first training sessions.

He drew the picture shown in Figure 2.3, which he described at the time as a quality circle. Although this is not the same concept of quality circles as was developed extensively later by the Japanese, Deming was making the point that people can only give of their best at work if they can balance the effort and satisfaction there with that at home, and vice versa. A problem either at work or at home affects the other part of the quality circle. Unfortunately, this concept is one that the Japanese soon forgot; one major problem in Japan is the very long hours that employees are expected to give to their companies and the great sacrifice expected to their home lives. Despite Japan's enormous industrial success, this does not detract from the fundamental truth in Dr. Deming's argument. Japan might have been even more successful if it had listened to this point, although by listening very carefully to Deming's other principles (which the USA did not) Japan has come to represent the epitome of high quality.

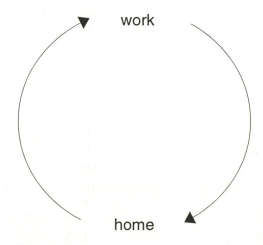

work

home

Figure 2.3   The quality circle

How, therefore, can you balance your life better in terms of your efforts at work and your family, hobby, commitments and interests? The answers to this question are all important parts of finding out your real ambitions.

It may be, too, that your ambitions are very long term and even seem ridiculously far fetched, but these ambitions may well be much easier to achieve than you might think. If you have a clear understanding of where you wish to get to, then very often all the factors you need for success somehow begin to fall into place. Jimmy Carter, for example, had the opportunity to meet one or two US presidents and decided that he was at least as able as them to be president of the United States—and eventually was! Similarly, Bill Clinton met President Kennedy when he was young and set his sights, too, on becoming president.

W. Edwards Deming, whom we have just mentioned as an eminent authority on quality, begin his working life as a janitor employed by his university.

2.3 Management = coaching

One of the questions I have been most frequently asked when recruiting students and graduates over the years has been: 'How can I become a manager?'.

Since many of you will also have, or have had, this ambition, consider an alternative question:

'What do you mean by the word *manager*?'.

Take a sheet of paper and write down some words which describe the key responsibilities of a manager. Think back, if you already have some experience of work, of the main criticisms that you have made of your own managers. Was it their inability to develop your abilities and career as much as you would have wished? Is this ability to develop staff, therefore, not the most important role of a manager?

In Figure 2.4 we can see the organisation pyramid as it is usually shown in organisation charts. The managing director sits at the top, with the production, clerical, development, marketing, etc., staff at the bottom. As the introduction to my book, 'Japan's winning margins—management, training and education'[1] says in its preface:

'On 11 February 1981 I attended a prestige lecture entitled 'Marketing, management, and motivation' given to the Institution of Electrical Engineers by Akio Morita, chairman of Sony. "Peter Drucker," he said, "recently wrote that he is unable to understand why Japanese industry is so successful. In particular, he seems puzzled by the fact that, in comparison with their North American and European equivalents, Japanese managers appear so unimpressive when you meet them. Peter Drucker's problem is this: he does not understand what management is about. In Japan a manager's role is very simple; it is to develop the skills of his staff so that they can find better ways of satisfying the customers."

'At that moment began my fascination with Japan. For many years I had been frustrated by the incompetent man management in British industry. None of the many training courses I had attended seemed to provide any real solution. Here it seemed to me might be the beginnings of a set of answers.'

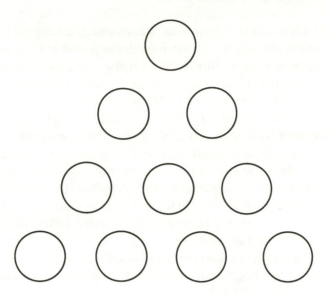

Figure 2.4   *The normal western concept of management*

In Japan, when a manager or supervisor manages to coach and develop one of his staff so well that that person is promoted to a more senior position than the manager himself, there is a special ceremony. The guest of honour is not the person promoted, but rather the manager who has demonstrated the epitome of good Japanese management.

The secret of Japan's successes is that, on average, it has the most competent employees in the world. And this is because the prime role of a manager in a Japanese company is the development of his staff. This ability to coach and develop is the primary skill required for someone to be considered for appointment to a supervisory or management post. Furthermore, it is on the basis of further developing these coaching skills that the initial training of managers is based. Even more importantly, it is the key criterion used to determine how a manager should be appraised, rewarded and promoted. Ask yourself whether your organisation thinks—and, more importantly, behaves—in this way?

If, then, the key skill that you will need as a manager is that of coaching, how can you develop and enhance this skill? First, both you and your organisation need to value this skill. Secondly, Chapter 7 will provide you with practical advice on how to develop it even further.

2.4 The learning organisation

As Figure 2.4 illustrates, the normal way in which an organisation chart is shown is with the chief executive at the top and the lowest ranking employees at the bottom. But that is not the way in which Akio Morita described the organisation of Sony and other Japanese companies when he gave that lecture to the IEE in February 1981.

Instead, Figure 2.5 shows the way in which a Japanese organisation works. The most important people to the organisation are actually outside it—the clients and customers. The prime role of the organisation is then to become a learning organisation—with the major aim of developing the most competent people possible in order to satisfy the clients. In fact, one way in which Japanese organisations have been described is as learning machines—and indeed this description is often applied by the Japanese themselves.

Competence development of individuals is, then, seen as being an integral part of the organisation's strategy, and not as an add-on. The

customers

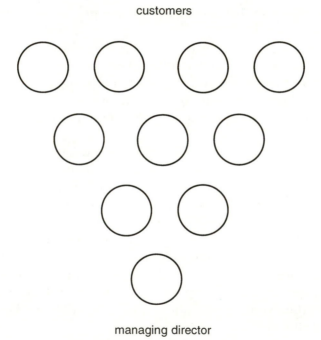

managing director

Figure 2.5   The concept of upside down management

career development of employees is similarly seen as both long term and integrated. For example, it is common in British organisations to find that the directors and chief executive have had narrow areas of experience throughout their careers. Not so in a Japanese company; invariably the top people will have had a highly systematic career progression so that, over a period of many years, they will have obtained detailed practical experience of all aspects of their organisation's business. They are then in a far better position to make the best strategic decisions—which on the whole most Japanese companies do far better than their British competitors.

Nothing, however, stops any organisation anywhere in the world applying these simple principles, and becoming world class. Nothing, perhaps, except the narrowmindedness and stupidity of their directors and chief executive!

As the quote by Henry Ford at the beginning of the chapter indicates, success is to a large degree dependent on your frame of mind. Achieving your aims, including becoming world class, is all about believing that you can do it.

Chapter 3
The responsibility of the individual

'No-one knows what he can't do until he tries'—Publilius Syrius, 42 BC

As Publilius Syrius noted two millenia ago, most of us are limited in our careers much more by the barriers in our minds than by any external barriers. And yet how many of us, at some stage or other, blame our bosses or our company for limiting our career development?

The hard, but highly stimulating, fact is that we all need to think through very clearly what it is that we want to do in our lives—and then go for it! You will be amazed, almost certainly, by how things then fall into place.

In my own case, I set myself the ridiculous ambition of becoming the world's greatest expert in CPD. Sitting there in Coventry, surrounded by some rather mediocre company management, it might have seemed a pretty crazy objective to many. There were some things that I did when the opportunity arose, such as becoming chairman of the IEE's career development (now professional development) committee, joining the International Association for Continuing Engineering Education and becoming a Council and Executive Committee member and then Chairman of its Professional Development Committee, and speaking at many conferences around the world.

My really big opportunity, however, came through the highly painful process of being tipped out of my then company by a new boss who simply did not want me in his team—which after even 27 years' loyal service can and will increasingly happen to many of us. After what I had considered to be a highly successful career in the company, this was extremely difficult to adjust to—even if it did cost the company concerned a large sum of money in compensation! But with hindsight, I realised that I had long wanted to move into the world of consultancy and to own my own business. I had, however, neither sat down and really thought through what I wanted to do, nor had the courage to make the jump.

By taking the opportunity of setting up my own consultancy business, I found enormous satisfaction in a number of new areas. For a start I persuaded The Engineering Council to let me author

'Continuing professional development—the practical guide to good practice', designed as a very pickupable and easy to read guide to CPD, and to give me a contract to edit 'CPD Link', its CPD newsletter.

I also obtained contracts from both Oxford University Press and McGraw-Hill to co-author two management books[1,2]. In addition, I obtained consultancy work in various areas of CPD, including running workshops in Hong Kong at the invitation of the IEE's Hong Kong Centre. Then, in 1996, I was asked to go out to Hong Kong to help the Hong Kong Institution of Engineers to launch its CPD scheme.

Now, with a very clear idea of how I want to develop my future career, I increasingly find that things fall into place surprisingly fast. With hindsight, being forced to leave my relatively comfortable job in a major company was the best possible thing that could have happened to me, despite the great pain and worry at the time.

I feel it is worth relating the above story of what happened to me because I increasingly come across more and more people in engineering who find themselves in very similar situations to the one in which I found myself. Realising it might well happen to you is also a strong motivator for placing great emphasis on your CPD.

The simple message, therefore, is that you should ask yourself how you want to develop your career:

- over the next year;
- over the next five years;
- over the next ten years.

What opportunities are really open to you that you have not properly identified? What would you do if your boss told you tomorrow that you are no longer needed in his team? What are your strengths? Who might be interested in making best use of them? What is the real value of the experience that you have had up to now?

Two books which you may well find it well worth reading are both by Professor Charles Handy: 'The age of unreason'[3] and 'The empty raincoat'[4]. These emphasise the need to develop a personal portfolio of competencies and to be aware of the possibility of having several different jobs at any one time in one's career.

3.1 Some key steps to take

There are a number of key steps that you need to take to maximise your continuing personal development. These will be described in more detail in future chapters, but some of them are:

- structure your competence development to maximise the benefits from learning activities;
- develop a clear understanding, in order to do this, of key competencies;
- use a personal development plan to achieve this;
- find a mentor, not just for initial structured formation training, but for the rest of your career. This mentor might change over the course of time, but should be independent of your managers. Later on in your career you, in turn, should mentor one or more other people;
- maximise every opportunity to learn. This not only means linking such learning to one's own key competencies and career objectives but, just as importantly, maximising ways of sharing such learning with others. If you do this, others will trust and respect you—and in return they will be prepared to share their learning with you;
- develop as wide a network of contacts as possible.

Now, as a way of demonstrating the opportunities for CPD that might be open to you, let us look at a number of case studies of individuals (previously featured in 'CPD Link').

3.1.1 Sally Martin—instrument engineer, Shell Refinery, Stanlow

'When I joined Shell as a graduate in 1987 I often felt frustrated by the constraints imposed by the IEE's structured training requirements. At the time I wanted to remain longer in each department to see the projects through, but my mentor insisted that I stuck to the programme. It has only been during the last two years that I have seen the real merits of this. First, the job rotation provided me with broad technical experience, secondly, it gave me the time to get an overview of the company and, thirdly, I developed a very significant network; this was enhanced by a 16 week technical training course in Holland which forced me to spend time with a range of different people.

'Learning is not only about knowing the answers, but also about knowing where to get them. It is very useful, for example, to have the telephone numbers of people to contact when you have problems.

'Shell has given me many superb opportunities for learning, including the chance to spend six months commissioning a new catalytic cracker, as part of a commissioning team, early on in my career. It was a high profile, fast moving, high responsibility job and gave me the chance to learn more in a short period of time than I will probably ever be able to again. Other opportunities have included helping with graduate recruitment interviews, working in the personnel department for one day a week on two major projects, giving talks in schools and editing a company newsletter.

'The company has worked very hard with me to identify my ultimate career goals. As a result I feel that I am managing my career and am in control of it. It is a very structured approach, just like my initial IEE training.

'My ambitions are managerial, rather than technical. One by-product of moving around is that you learn to deal with people. Too many technical people in this country are not trained to be managers, but nevertheless end up in management without having learned to deal with people. One of the key things I have learnt is that being an engineer is much more than being a technical expert.'

3.1.2 John Hawkins, CEng—consultant

'CPD has for me been a bridge between two careers—thirty plus years working for a multinational company, and running my own consultancy coaching and mentoring senior people in industry, healthcare and the professions.

'With Alcan Aluminium, based in the UK, I started in sales development on new product and process management, later setting up the first marketing department after spending time getting my MBA in Switzerland.

'I went into general management, first in consumer and packaging, then in speciality products and finally in aerospace and defence materials. As divisional managing director, and part of Alcan's management team, takeovers, startups and turnarounds were the norm. Inevitably there were many rationalisations, plant closures and redundancies. Whatever satisfaction we obtained from the bottom line profit, growth in exports, etc., I rated equally important the growth and development of the people coming up through the organisation. A coaching style of management meant it was exceptional to let go a senior manager whose job had disappeared, or who proved to be a square peg. More often, through career development, a new slot— maybe a totally new direction—was found to the advantage of the individual and the organisation.

'I had served on the continuing education and training (CET) committees of The Engineering Council and the Institute of Metals, so I was happy to organise production of promotional material for the launch of The Engineering Council's CPD programme, notably 'CPD—the practical guide to good practice'.

'Acting as specialist advisor to the House of Commons Select Committee—Trade and Industry (reporting on the competitiveness of UK manufacturing industry), and tutoring manufacturing strategy for the Open Business School MBA, enables me to take a view of best practice in the UK for people development. I am convinced of the importance of developing a culture which encourages self-improvement for all employees, and recognising the mutual benefits of such an attitude—namely the importance of CPD.

'Today an individual with a lifetime career with one employer is an endangered species. More often that hard to get job will be a short-term contract—expecting you to assume self-employed status, taking care of your own tax, insurance and pension.

'So it only makes sense to take charge of your own career development.'

3.1.3 Mark Meffan, CEng—project manager, Sir Alexander Gibbs & Partners

'I originally joined the Property Services Agency as a civil engineering graduate and followed the Institution of Civil Engineers' training scheme. This provided a very proactive approach to my training needs. There was a nice balance between meeting the company's needs and my own, as I was fully consulted on where I wanted to take my career through appraisals.

'Since my interest was mainly on the management side, I took on more demanding management positions, as well as attending a series of management courses. In 1991 I spent six months in the Falklands in charge of building a new road from the airport at Mount Pleasant through to Goose Green. This gave me the opportunity to be self sufficient and the chance to learn by experimentation. We were living well out in the sticks, ten miles from the nearest habitation; I had to keep my team fed, warmed and even entertained in the evening.

'Before I took redundancy from the PSA in April 1993, I had become a part-time tutor on the Science and Engineering Research Council's graduate school programme, tutoring groups of ten for three weeks. I continued this while I was looking for a new job—a task which I approached by producing a structured programme of how I would contact organisations to apply for jobs.

'I joined Sir Alexander Gibbs in September 1993 and went on a number of very useful orientation courses. I have now achieved a balance in my experience between the theoretical and the practical and am able to realise my career ambitions as a project manager.'

3.1.4 Aileen Aulds—senior communications and electronics officer, Metropolitan Police Service

'I graduated in 1983 with an honours degree in physics from Leicester University. At the time jobs in technology were difficult to find, so I spent three years as a chain sales floor manager. This experience of people management certainly helped me at the interview for the Metropolitan Police Service (MPS) two year graduate engineer training scheme, which I joined in 1986.

'On completion of training, I was assigned to the X25 wide area data network planning section. I was surrounded by many experienced people, particularly incorporated engineers, and to me an important aspect of CPD is obtaining inputs from everyone around. This was extremely useful as I came from a nonengineering background.

'With a view to improving my technical expertise, in 1989 I requested, and was granted, day release to study for an MSc in information systems engineering at South Bank Polytechnic, which inevitably involved a great deal of work in my own time. During my second year on the course, I was one of 13 engineers selected to participate in a management development programme being run by The Industrial Society on behalf of MPS. This consisted of a series of training seminars, discussions and in-company projects over a period of nine months. I gained a lot of confidence, particularly from being able to discuss my ideas with others in the privacy of the seminar room.

'The experience of overlapping these two courses plus day to day work demands was invaluable. It was a time management and organisational exercise, putting me under a lot of pressure. There was always something to do on the train!

'Recently I was promoted to senior engineer in charge of the local area networks projects section, which provides all local area network data communication requirements for the MPS. To keep up to date, I keep an eye on the technical press and publications relevant to my field, while being selective about what to read in detail.

'I will be using one of the IEE's professional development records to help me to structure my learning even more effectively. I use all the contacts I have available to increase my rate of learning and I believe that everyone needs to be aware of their own development needs and be prepared to drive their own training. That is what CPD should really be about.'

3.1.5 Bob Hawkins—development manager, ICL, Manchester

'I started a masters degree in business management at Manchester Business School in 1989. This was a very intensive course, with 55 days of residential periods split into six full weeks and six long weekends spread over the first 15 months. There were assignments all the way through.

'I had to study for ten to 15 hours per week over the two years duration, and I also tried to spend time reading newspapers. It is also important to read around the subject, not just text books, but also business publications, for example 'The Economist' and 'The Harvard business review'.

'In a functional organisation you can get a narrow view of the world. I did this course to prepare for the future by getting a general business education and training, and this has given me more rounded skills and awareness.

'I think it is important that people manage their own development and do not expect their employers to do it for them. Within ICL, however, there is considerable support for personal development, which is very important for those undertaking such a task.'

3.1.6 Raiya Fells—regulatory and certification co-ordinator, Alba Project, Chevron UK Ltd

'In October 1991 I became one of 20 full time quality coaches in Chevron.

'My advice to anyone wanting to develop their career is to broaden their horizons in terms of supervisory experience, contracts, planning, budgets and cost control through small project engineering assignments where possible. Chevron provides an average of five to seven days of off the job training each year.

'I am an associate of the Institute of Metals, and this has been very useful as a vehicle for eventual chartered status. Their technical publications are also very good.

'I am keen to help women to develop their careers in engineering. Thirteen per cent of the staff in supervisory and professional grades in Chevron UK are women. When I was in Aberdeen I helped the SATRO by giving talks in schools to 11–14 year old pupils about careers in engineering.'

3.1.7 Dr Richard Pike, CEng—director general, Institution of Mechanical Engineers

'Continuing professional development involves regularly taking stock of one's career to date, and planning the acquisition of skills and qualifications to achieve future ambitions. This will involve a range of programmes from company-specific training to the development of more general attributes that make one more transferable or promotable within or across industries, with the balance depending on economic cicumstances, job security and the challenge of new opportunities.

'I began my career with BP, and adopted a practice of annually revising my curriculum vitae to put achievements, experience and priorities into perspective. Wherever possible, major investigations and reviews were summarised into reports, which ultimately formed the basis of my submission for achieving corporate membership of two major engineering institutions.

'With both the boom and the uncertainty of the oil and petrochemical industry in the period 1968-93, I sought to maximise my experience and qualifications, and altogether held technical and commercial positions in ten different locations in the UK and Japan, achieving qualifications in four foreign languages, and being in the fortunate position to obtain CEng, FIMechE, FIChemE, FInstPet to add to my academically-based MA, PhD. My most recent positions include manager technical, Sullom Voe Terminal, Shetland and president, BP Chemicals, Japan.

'By early 1993 I recognised the risks of being in an industry in possible long-term decline, and resigned from BP to become director general of the Institution of Mechanical Engineers. I would like to think that my experience and attention to CPD were contributory factors in my appointment and the support which I have received since.'

3.1.8 Alan Robson—factory manager, Boots Pharmaceuticals

'After starting my career as a mechanical technician apprentice with Vickers, I then did a mechanical engineering degree and joined Boots in 1978 as a graduate engineer.

'I have held a variety of positions in the company, including production engineer, project engineer, factory engineer, works engineer, engineering manager and chief engineer; in the last instance I was responsible for production facilities at a number of sites across the country.

'Six months ago I had a complete functional career change, moving to be factory manager in the company's largest factory; this has 500 staff and outputs over £100 million of products each year.

'During my career I have had to make major capital decisions. I therefore did the certified diploma in accounting and finance, as well as various courses at Ashridge and Henley. I also read up a lot on general management issues such as in the IMechE literature and 'The Harvard business review'.

'I have chaired events for the IMechE, as well as taking a wide interest in the development of young engineers. I am now a principal industrial mentor, mentoring four undergraduates and graduates.'

3.1.9 Peter Bell, CEng—entrepreneur

'Many of our great 19th century engineers became famous because they were entrepreneurs. Although we live in a very different cultural and business climate today, I believe that there are equally great opportunities for engineers to become entrepreneurs in manufacturing, product design and development and engineering consultancy. Indeed, a knowledge and experience of engineering, coupled with continuing professional development, is an ideal background for the would-be entrepreneur.

'Looking back on my own career over the past 40 years, I have progressed from the drawing board to engineering management in a variety of jobs in different industries, with three years overseas. I joined the Institution of Mechanical Engineers in my early 30s, becoming a fellow in my early 40s, but at the age of 44, became redundant, following a takeover. Being an individualist, this provided the opportunity and the motivation to take the plunge as an entrepreneur. Recognising the need for further training in business management, I used part of my severance pay to acquire a knowledge of accountancy and marketing at a leading management school. Through various contacts, I then found a niche in the motor accessory business which, in turn, led me to design several protected products to fill some of the gaps that I saw in the marketplace.

'I started manufacturing and, despite various crises including three-day weeks and power and fuel shortages, my company, Bell Products, achieved a turnover in excess of £2 million. Later I established a second, equally successful, business for manufacturing display packaging. More recently, I have established a classic car restoration business and become chairman of a small electronics firm.

'My engineering training and broad experience provided me with an innovative approach to problems, confidence in dealing with people and a clear aim in setting and achieving targets. Admittedly, a lot of hard work is involved in creating and running a successful business, and determination and an element of luck certainly play their part.

'I suggest that those engineers in mid career who may find themselves faced with a lack of opportunity for advancement, job frustration or redundancy should look upon this as a window of opportunity to become an engineering entrepreneur. In my experience, this will provide a challenging and exciting career offering tremendous satisfaction in creating employment for other people. If you are manufacturing your own products, you will have the added thrill of seeing these products on sale, perhaps on a worldwide basis. You will also make a positive contribution to the rejuvenation of the wealth creating sector. Being your own boss undoubtedly has other attractions, of course, in terms of financial rewards, status, opportunities for travel, meeting interesting people, influencing affairs, etc.

'The time to start thinking about this is now. By taking a positive approach to CPD, both technically and managerially, you will be well prepared when the opportunity arises.'

3.1.10 Francis Frampton—jack of all trades

'My multidisciplinary interest in engineering caused me to join three rather different professional institutions—the Institution of Engineering Designers, the Institution of Agricultural Engineers and the Institute of Data Processing Managers.

'In 1954 I started a new nursery, but no firm would make a glasshouse to my specification. So I got out my Oxford University Air Squadron maths and physics books and designed and built my own. It reduced the cost of aluminium glasshouses by a third. Although I was a horticulturist, I eventually built up a multimillion pound metal bashing factory and construction firm, building glasshouses from Canada to Israel. One unique botanic garden project was described in the finale of an eight part architectural series on TV as 'the finest example in Britain of the use of modern materials in the creation of an environment'.

'I was gazumped in a family share deal and left horticulture abruptly! So I worked as a consultant in a huge organisation on design and procurement within tight quality constraints. I gave a presentation on 'How the Japanese gain competitive advantage when making motor cars' and am now applying their techniques to other industries.

'One client supplies emergency spares to superyachts worldwide. He has more than doubled the proportion who receive deliveries within three hours of the minimum theoretical time. Without extra staff, he also doubled his turnover and trebled his profit last year.

'I am working with the CEO of a county Training and Enterprise Council on an autumn programme aimed at helping firms to become world class manufacturers. Yet another project is aimed at helping a small industry to become world class; when I tell the staff that my role is to help them gradually become world class, their eyes light up!

'I believe I bring honest ignorance and a fresh mind to problems, maximising each individual's potential in increasing competitiveness. How many of us, I wonder, ever really achieve our full potential?'

3.1.11 Diane Davy, IEng—assistant secretary, Institution of Mechanical Incorporated Engineers

'I have become very committed to CPD and the use of a personal development planning document because I have used the process and it has worked for me.

'I believe that practising CPD enables you to take control of your own future in a subtle way. Despite the accelerating rate of change around us, it is all too easy to remain static in your skills—and thus lose control of your own self development. CPD lets you take charge and perceive possibilities by looking forward, so that you develop the skills to suit each envisaged scenario; then you are hopefully not left behind.

'I originally trained in electromechanical switchgear and small mechanisms. Then I took a career break to have my family, but was determined to return to the engineering career that I loved. I had difficulty finding the part-time work that I wanted, but was eventually lucky enough to get a job as a laboratory assistant, in materials testing, in a research laboratory. Over ten years my working hours gradually increased to full time, including spending half my time in each of two departments. Nevertheless I became bored and wanted the stimulus of something more challenging.

'I then moved to work at The Engineering Council for five years. With the change in my working environment I desperately felt the need to upgrade my skills. I therefore spent time one weekend completing one of The Engineering Council's career manager documents. I found that this both highlighted areas of shortfall in my skills and, surprisingly, gave me more confidence in the skills that I already had. I still regularly fill in my career manager, keeping a record of my personal development.

'A year ago I moved to my present job where I have found it immensely valuable to redo the career manager self analysis. One of my tasks has been to launch the Institution's CPD record book which aims to combine the best features of such documents from The Engineering Council and other institutions. This is now selling steadily and will surely benefit our Institutions' members every bit as much as I have benefited from using such a document for the past six years.'

3.1.12 Christopher Davies, CEng—managing director, Industrial & Municipal Engineering Services Ltd

'I see CPD as being about a need to obtain broad experience. In my case, this has been largely overseas and, unfortunately, I find great difficulty these days in recruiting contract engineers with the necessary breadth of international experience.

'Setting out as a regular naval officer, and being invalided out rather early in this career, I became a civil engineer via a spell on the factory floor in the aircraft industry, Cambridge University and service in a firm of London-based consulting engineers.

'There is no doubt that, in this branch of engineering, one must be happy and confident operating alone. I was lucky to have appointments in the Middle East, Far East, briefly behind the Iron Curtain and in Latin America. All were interesting and valuable experiences, but probably the most unusual was being in Cuba up to, and during, the Castro revolution. Our client was the Bank of Cuba and when the head of the bank resigned in exasperation, Che Guevara was appointed in his place and therefore became my boss! He also showed unfailing courtesy and good manners in all my contacts with him.

'By the late 1960s I had gone private, and had developed a design and build business in water and public health engineering. The Arabian Gulf proved to be the best stamping ground—it has been now for over 28 years—bringing government and oil company work.

'I became more mechanically and electrically orientated, with an interest in the pathogenic side of the public health work, as well, and by 1990 the company was getting work from the UK water utilities.

'I had a very interesting spell in the 1980s in Paraguay on secondment to the UK's Overseas Development Administration, advising on pollution control generally and helping with the preliminary drafting of the technical elements of legislation for dealing with the problems arising from tanneries and abattoirs.

'Today, over normal retirement age, I am back in a consulting and advisory role, but still very active and acting, among other duties, as the European director for a firm of Australian international consulting engineers, with which I have worked on a number of occasions over the last 25 years.

'I don't see the work load lessening very much, but it will no longer include the hassle inevitably associated with contracting.'

3.1.13 Isabelle Jenkins—Royal Bank of Scotland

'I joined the bank's information technology department with an MEng in electrical engineering with business studies from Imperial College, London, and looked to use both my technical and business skills.

'Like every member of staff within the technology department, I draw up an individual development plan (IDP) with my manager. This is negotiated and revised during the year, so that you look at both your current job and where you want your career to go and identify the skills needed. You then draw up a plan to develop those skills; this involves far more than simply identifying courses, but also includes targeting the specific broadening experience needed. This may include work shadowing, open learning or external courses. Some of this is carried out in work time and some outside work. New initiatives, such as monthly talks by members of staff, from both inside and outside technology, are being pioneered. These tend to be informal and interesting sessions when one can build up a picture of the bank's business as a whole.

'As part of my IDP for this year, my team and I are participating in a management game where we are running a manufacturing company and have to make decisions on costing, production and competitive strategy. This, in turn, should enable us to gain a better understanding of the pressures experienced by our customers.

'Last year the bank also let me take six weeks off to work as a volunteer for a conservation charity in Indonesia, where I was a staff assistant on an expedition of 50 people in the rainforest, and also worked at an orang-utan rehabilitation centre. This experience enabled me to develop both my interpersonal and leadership skills, and to bring these benefits into my job.'

3.1.14 Bill Hoult—chartered surveyor

'Most professional organisations now require some form of evidence of CPD from members and it is usually quite a small amount each year.

'My own main professional institute, the Royal Institute of Chartered Surveyors, requires evidence of at least 60 hours in a three-year period—a quite ludicrously small amount, and one must doubt the ability of any professional who claims that he or she is unable to meet such a target.

'I have kept a record of all my CPD over the last three or four years for my own professional interest since the RICS requires it to be earned on an hourly basis.

'What do I include? The following:

1 Formal training courses, and if a CPD certificate is available then I take it. If I am involved in the presentation, then I add additional time to allow for my own preparation (which is basically structured study). Last year, for example, I gave a number of seminars to NHS estates on fire code documents for the Institute of Hospital Engineers (IHE) and as a result joined the IHE as a fellow.
2 Meetings where, for example, the approach to a particular project is discussed with fellow professionals of the same or a different discipline. (In my view, this is the most effective CPD of all.)
3 Private study, provided that it is in a structured sense. (In my case, recently coming to terms with the construction design and management regulations and related approved codes of practice.)
4 Technical exhibitions, health and safety, fire, security, etc.
5 In-house presentations.
6 Presentations on business administration (bookkeeping, marketing, etc.). These may seem unrelated, but successful running of a business is all part of professional competence.
7 Management presentations such as 'The Disney experience' or a lecture on 'Gestalt' (a management philosophy). These are, of course, directly related to item six.
8 I am a councillor for a local authority, and as part of that duty I hear presentations on such subjects as planning applications, the proposed local plan for the area, installation of cable networks, etc., all of which I add to my CPD records. I would point out that most Council meetings where these presentations are made are open to the public free of charge.
9 Recently there was a series of programmes on the television on architecture; I missed all of them, but would have included them in my CPD if viewed.
10 More contentiously, I would add visits to such exhibitions as the visitor centre at Sellafield and the Channel Tunnel exhibition. These keep one abreast of the latest developments in technology.

'I normally keep some literature relating to the subject even when CPD certificates are available.

'In short, I reject the narrow view of continuing professional development that says it must relate to my specific work or indeed profession. Unfortunately, this narrow view has been compounded by some institutions who appear to be insisting on points systems backed up by certificates to be issued only by accredited training organisations. This approach ignores the experience obtainable in the wider world of information and is short sighted. (Try getting a CPD certificate out of the BBC!)

'Obviously taking the wider view, as I do, means that my CPD is easily in credit, so any questionable experience may be disregarded without loss. It also alerts one to the value of the differing resources available to us in the wider world of information and encourages the use of them.'

3.1.15 Clare Gallagher, CEng—personnel manager, Zeneca

'My really powerful learning experiences have come about when I have been thrown in at the deep end. For example, after graduating in mechanical engineering, I was appointed as a technical support engineer. In some respects it was a painful experience, but I was able to learn quickly. Later as an operations manager in the early 1980s I was fulfilling a role new to the company, combining production and engineering into one. I certainly learnt a lot from that. I was managing a plant which had to operate 24 hours a day, seven days a week, on a continuous basis; there were technical challenges with many interlinked processes and the management of people on different shifts.

'Then I went to Brazil for two years, where my work included implementing a quality system. Some of my staff only spoke Brazilian Portuguese, so I had to learn to speak that language fast!

'In my present job I have been able to support the strong emphasis in the company on mentoring, coaching, on the job experience, professional competencies and personal development plans. We have a 'Guide for mentors' which is widely used, as well as mentoring workshops to keep everyone up to date with current thinking. The focus is on the coaching role between line managers and individuals. We are currently running coaching skills workshops to continue to build people's skills in this key area; we are starting with the managers, but plan to go much further, giving employees the skills to coach others in the organisation. We have also built on the considerable work on identifying individuals' key competencies that we did when we were ICI. In addition, everyone has a personal development plan as part of their personal review form, and this is reviewed informally on a regular basis and formally once a year.

'It is vitally important for everyone to carry on learning as they go, otherwise the world will pass them by. The company needs to provide strong support, but it must be driven by individuals who take responsibility for their own CPD.'

3.1.16 Dr Yvonne Barton, MBE, CEng—head of gas sales and commercial negotiations, British Gas Exploration & Production

'My current job involves me in gas production operations as far and wide as Thailand, Trinidad, Tunisia and Teeside. In addition, I also head the company's acquisition strategy in Europe and South America, applying my engineering discipline to the financial world. Gas marketing requires a complete knowledge of the technical side, and one of my roles has involved leading a marketing campaign for a two trillion cubic feet gas field to be sold to the major energy companies. Throughout my roles I have retained my involvement in corporate finance. World energy markets are changing every day; in particular, the former Soviet Union has opened up and my trip to Moscow this year was a fascinating insight into the world's biggest energy company.

'My advice to all engineers, therefore, is to see CPD as giving the broadest possible expertise for dealing with the widest range of challenges which may be encountered.

'When, after research at Cambridge, I joined BP for my first job, the company had a civil engineering department which got involved in all aspects of marine construction, including site surveys, design, installation and research. It is an indication of how things have changed in the oil industry that such a department would now be regarded as a luxury, but it was a wonderful place for a young engineer to start and an opportunity to see the world. On site at the Wytch Farm Oil Field, gaining the necessary wet wellingtons experience for my chartered status, I grew fascinated by exploration, and saw that, whereas the names may have been different, the equations were just the same. My background in computer simulation enabled me to be posted to the reservoir engineering department, studying oil and gas flows through underground reservoirs. A posting to Aberdeen ensued—on the Magnus Field—where I was doing a special research project testing well productivity offshore. This job included protracted periods offshore and long helicopter flights to the most northerly platform in the UK; all sailors knit, so this was a golden opportunity to run up a few jumpers.

'Networking with landbound friends down south, essential when offshore, I heard of a new opportunity at British Gas. Following privatisation and the removal of its oil assets to form Enterprise Oil, British Gas had made a number of acquisitions internationally and needed people with overseas oil experience. I became head of mergers and acquisitions, which involved running a large multidisciplinary team covering international law, corporate finance, taxation, marketing, geology and the other technical sciences.

'As an engineer I have had the mathematical ability to understand the tasks ahead; as an energy specialist I have been able to direct and motivate the various elements towards a common goal—project team leadership.

'If I can attribute my success to any key factors, then there are two in particular. First, being prepared to take up any challenge which came along. And, second, being determined to prove people wrong when they have told me that I could not do something!'

3.1.17 David Mason, CEng—managing director, Mason Communications

'I formed Mason Communications in December 1992, together with three fellow directors. We started with 12 people and now have a team of over 70, of whom 15 are chartered engineers. We are an independent telecommunications, IT and security consultancy providing strategic planning, system design and project management services. Our customers are presently in the UK, USA, Spain, Hungary, France and Sierra Leone.

'As a team of directors, we have always placed great emphasis on the importance of professional team training and this was the driving force behind the development of our graduate training scheme and our successful accreditation by the IEE in 1994.

'In the rapidly changing technology and business environment in which we operate, it is vital to the success of our business that we continually train and develop our staff, because they are our best asset. We do this by regularly reviewing the training needs of our people and encouraging them to come forward with their own ideas on their future development. Having identified the needs, we use a mixture of internal and external training, together with on the job training, to achieve the objectives. We are sponsoring two people on part time business studies degrees, one person on a part time MBA and two of our engineers, who want to achieve chartered status, on Open University courses. In addition, all four directors are about to start the Institute of Directors' diploma course.

'We are participating in the Investors in People scheme and in addition, our quality assurance system is accredited by the BSI to BS EN ISO 9001.

'One of the most powerful learning experiences earlier in my career was working in project management, where I learnt about the commercial and financial aspects of business, controlling programmes and leading teams of people. These skills were invaluable additions to my engineering background. Later I ran a business within a larger business, with overall responsibility for profits, satisfying clients and obtaining new business. After setting up my own business three years ago, I faced new challenges such as recruiting new staff and building the team. We are developing new alliances and opening new offices; as part of this we are taking part in the DTI's Initiative Europe scheme and I have therefore been learning a new language for the past year.

'I believe that CPD is not only important to the individual, but is also vital to the success of companies in today's rapidly changing business environment.'

3.1.18 Neil Hardiman, IEng

'What should be eligible for CPD points? Engineering, management and related formal courses, evening lectures, plant visits? Yes—it is clear that these should qualify, but what of private study? Some institutions allow this to be claimed towards CPD, but others do not.

'In my view, private study must be part of one's CPD. Time spent studying a relevant book, television programme, British Standard or professional journal on an engineering or related subject is likely to be of as much value as attending a lecture or course, and probably far cheaper and more convenient. While a lecture, in my experience, is often a good introduction to a subject, the real learning comes in further private study.

'You may enjoy good health, have some or all of the costs of your courses paid, and a job and an income. I did too when, at 29 years of age, illness devastated my health and forced me to give up work in 1992; it simplified my career plan to the single word 'postponed'.

'Thanks to my Institution's (Plant Engineers) continuing professional development scheme, institution lectures and visits, local libraries, television and radio, and private study at all hours of the day and night, I now know more about engineering, and something about management.

'But how can I show that I have conducted this private study? On average I find that I need to produce at least one page of typed notes per hour of study, as a future source of reference, which my Institution has accepted as evidence of my record of CPD.

'CPD has been invaluable to me. It has relieved the boredom and frustration of illness, provided essential mental stimulation and a verified record of CPD for any interested future employer.'

3.2 Summary

I hope that you have found some useful ideas for your own use by reading these case studies.

Take a bit of time now to sit down with the sheets of paper and ideas that you have already put together as a result of reading the book so far. Review them and see how any ideas from these case studies might change or develop some of those plans.

I have deliberately included two entrepreneurial examples. In both cases, like my own, circumstances caused the individuals involved to leave secure employment with a large organisation, and in both cases this led to even greater opportunities. So, for any of you who have either already been made redundant—or who might fall victim to the new boss syndrome—or who might at some future date find these or similar events happening to them—things need be nothing like as bleak as you might first think.

As I am finalising the draft of this book, there is news of the death of Air Commodore Sir Frank Whittle at the age of 89. He was probably the greatest aeroengineer of the 20th century. When he had his moment of triumph on May 15 1941 when the jet propelled Gloster-Whittle E28/39 flew successfully for the first time from Cranwell, it was the result of overcoming years of obstruction from the authorities, as well as lack of funding for, and lack of faith in, his brilliant ideas. We can all learn from his great tenacity in the face of adversity. So often, great triumphs can arise for each of us if we have the courage and staying power to see our individual visions to their fulfilment.

The final case study is, however, perhaps the most inspiring. Bad health can strike any of us at any age. Would we rise to the challenge as well as Neil Hardiman has done and use CPD to cope with the problems? And for those of us fortunate enough not to have been so challenged, what are we doing to recognise the great advantages that we already have—such as good health—and to use those as a foundation for really challenging the world with our inherent abilities?

Chapter 4
How to identify your key competencies

'More new knowledge has been produced in the last 30 years than in the previous 5000'—Peter Large, author of 'The micro revolution revisited'

Wherever we are in engineering, we hardly need reminding of the extraordinary pace at which new knowledge and expertise is being produced around the world. It is all too easy to feel overwhelmed by it and to be tempted to abandon the effort of keeping up. Instead, surely, we should all be enormously excited by the opportunities that it opens up to us. How can we handle this accelerating change? James Appleberry, president of the American Association of State Colleges and Universities, for example, said at the First World Lifelong Learning Conference in Rome in November 1994 that 'one of our former US presidential cabinet members projected that by the year 2020 information will double every 73 days'.

One important way in which we can maintain our sanity, and structure our learning and career development effectively, is to have a very clear idea of our key competencies. Once we have done this, we can then concentrate on ensuring that we seek out new knowledge which maintains, raises and broadens these competencies. Otherwise, there is a danger that we are shooting blindly and randomly in our efforts to keep up to date. We can then review our competence development regularly, with a daily review of what we have learnt and how that relates to our key competencies and a review over the past few months and years to see what progress we have made. We can also structure our future competence and career development that much more effectively, including deciding how to take advantage of opportunities as they occur.

Even apparent personal disasters like redundancy and being evicted by a new boss become very much more manageable if you have a clear understanding of your key competencies. For a start, you will be much better placed to ensure that you move your career in the best possible direction. And in addition, if you are applying for another job, you are far more likely to impress a potential employer if you have a clear understanding of the key skills which you can bring to a new organisation; this is even more true if you can show a personal development planner that you have used to structure, record and target your competence development. Employers who know what is good for them will be looking very closely at both the learning processes and learning potential of interviewees. It is a great mistake, particularly in a fast changing environment, to look only at the existing abilities of a potential recruit—a mistake all too commonly made; it is far more important, as Japanese companies do, for example, to consider the ability of a potential recruit to learn and develop as fast as possible. Maintaining a personal development plan centred around your key competencies and your learning processes will almost certainly give you an important advantage over other candidates for a job.

The next chapter will show you how to develop and use a personal development plan but first, in this chapter, we will look at what a key competence is and how you should go about identifying your own set of competencies.

4.1 What is a competence?

The definition of a competence which I like is: 'A competence is an asset of an individual who only loans it to the employer.'

This puts the ownership of the competence very firmly on the shoulders of the individual. Only you can really maintain and develop your own set of competencies, but how do you go about identifying and describing them?

There is an important distinction to be made between a competence—which is something you can do—and a knowledge. One way in which we can look at the link between these two is by looking at Figure 4.1. The three important elements are:

● knowledge;
● skills;
● attitude.

As Figure 4.1 shows, knowledge is linked to our academic ability. You can test your own, or someone else's, knowledge by asking them to write down what they know about the subject in question. Knowledge can be accumulated from formal tuition, such as school, university or a course. It can also be accumulated through day to day experience.

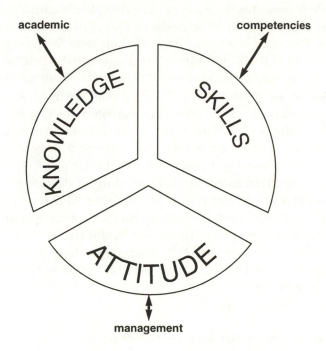

academic   competencies

management

Figure 4.1   The link between knowledge, skills, attitude and competencies

Skills are rather different. You cannot usually test someone's skills by asking them to write things down, unless they have a specific writing skill. You would test someone's skill by asking them to do it. Only by observing the successful performance of the task would you be reassured of competence. Imagine the furore there would be if a major rail accident was investigated, and the manager in charge of the person who turned out to be responsible for the accident said that the only evidence available beforehand that that person was competent was that he had written excellent essays about his skills. You only pass your driving test, for example, by showing an expert that you can actually drive safely, not solely by an academic written test.

The link between skills and competencies may be described as: 'A competence is the efficient, effective and proper application of skills based on appropriate knowledge.'

The third element in Figure 4.1 is attitude. It is only by having an appropriately positive attitude that skills can be efficiently, effectively and properly applied. No matter how good an individual's knowledge and how impressive their skills, you would be unwise to employ them or use their expertise unless you were reasonably confident about their attitude. Figure 4.1 links attitude to management because it is often the case that someone's approach to their work is strongly influenced by the ability of the organisation's management to provide motivation. I certainly remember a number of times in my own career when I did a less than perfect job because I was thoroughly disillusioned by either my own manager's, or the company management's, approach to me or my value system. On the other hand I can also remember many occasions on which I excelled, sometimes to a degree way above my expectations, when my manager inspired me in various ways.

In summary, therefore, knowledge, skills and attitude are different and distinct entities, but it is important that all three are present and interlinked. Skills must be underpinned by appropriate knowledge; for example, to pass your driving test in the UK, you not only need to show that you can drive safely by doing it in the presence of a driving examiner, but you also have to answer correctly some questions about the Highway Code.

4.2 Identifying your key competencies

In order to describe a key competence, you need to construct a sentence with an active verb, in other words a verb which describes an action. Some examples of these are shown in Figure 4.2.

Now make a first attempt at identifying those competencies which you believe are key to your present job. There should be at least half a dozen or so, and a maximum of 30. The lower the number that you can identify as being critical to your job, the better. You will need to closely and constantly monitor how you are maintaining, developing and broadening these competencies, which you will find very much easier with a small number.

Of course, we can all fill sheets of paper with lists of competencies, particularly if we are trying to impress others! But how many of these, you need to ask yourself, are really key. For example, if you are a secretary or a telephone salesperson, then your telephone skills may be critical. For most engineers, however, telephone skills, while important, are likely to be much less important than technical skills such as designing a product, or management skills such as coaching and developing staff.

It will take you some careful thought, and maybe a fair bit of time, to decide which your key competencies are. In order to identify them, try asking yourself the following type of questions:

Figure 4.2   Some examples of active verbs to be used for competency descriptions

- what was the key competence that I was using the last time I had to make an important decision, and the time before that, and so on?
- which competencies do I need to use in order to meet my appraisal objectives this year?
- if I were to be promoted, and had to select someone else to fill my present job, what are the vital skills that I would insist on when selecting my successor?
- in what areas do I regularly find myself being consulted by others inside or outside this organisation?

In each case, when you have identified a likely key competence, write it down using an active verb (such as those in Figure 4.2). However, do not write a sentence which uses the word 'knows'. For example: 'I know how a fork-lift truck works' does not describe a competence. The verb to 'know' is clearly associated with a knowledge, not competence. After all, the sentence could mean that you could describe the way to drive and operate a fork-lift truck. Alternatively, it could mean that you can write a detailed manual on how to build or maintain a fork-lift truck. On the other hand, sentences like: 'I am able to maintain a fork-lift truck', 'I am able to design a fork-lift truck' are much more specific about what it is that you can do.

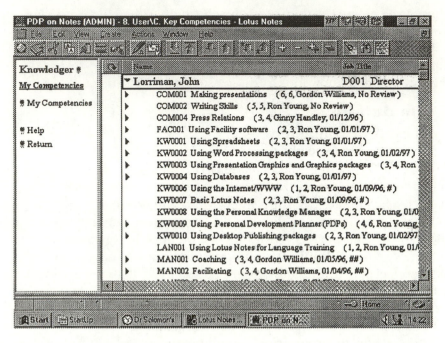

Figure 4.3   An example of the author's key competencies using Knowledge Associates'
PDP on Notes software

In my own case, for comparison, I have identified around 30 key competencies, some of which are shown in Figure 4.3. (The meaning of the data in the brackets to the right will be explained later in this chapter.)

In some instances, as you will see, I have used a sentence, such as 'using word processing packages'. In other cases I have just used a one word verb such as 'facilitating'. Make it as easy as possible for yourself, as long you are clear about what you are describing. Test each key competence description you write down by asking yourself: 'If a complete stranger came in, could I use this key competence description to explain in simple terms what the competence is about?'

One approach you can use is to divide your key competencies into two categories:

● technical key competencies—such as designing, marketing, manufacturing, using software, etc.
● nontechnical key competencies—for example, coaching and developing staff, interpersonal skills and the ability to make presentations.

You should ideally have a balance between these two sorts of competence. If you either have none, or only one or two, in either of these two categories, then ask yourself whether the balance is right in your case. Is there anything you should do to improve that balance?

Figure 4.3 includes other aspects of key competencies with levels, the name of the person who has agreed these levels and review dates. Therefore, let us now look at these aspects of the subject.

4.3 Levels of competencies

One of the frequent mistakes made is to assume that competencies are binary—in other words that someone either does or does not have a particular competence. But is the world really like that?

When we first start to develop a competence, our first efforts are usually less than perfect. With practice we improve our performance, but may well need support or supervision until we have mastered the skill concerned. Eventually we are so adept at it that we can safely perform the competence on our own. With yet more practice and experience, we become so competent that we are able to pass our skill on to others and tutor them through the learning cycle.

These stages in progression of competence can be translated into various levels. The framework that I prefer myself is:

Level 1: is aware of the key competence standard required;
Level 2: can frequently achieve the required standard of key competence;
Level 3: can consistently achieve the required level of key competence;
Level 4: can develop others in the key competence;
Level 5: can take corrective strategic action to redefine the key competence if necessary;
Level 6: is world class.

One of the reasons that I particularly like this framework is because it includes a level of competence which recognises the importance of passing skills on to others. After all, learning in any organisation needs to be a two way process. Those who know or can do, need to pass their knowledge and skills on to those who need to know or do.

Another reason for my own preference for this framework is that it also recognises that changing technology and situations may require a review of the definition and relevance of the competency. Only a real expert in the area is properly qualified to undertake such a review, and again I therefore believe this needs to be recognised in the framework as Level 5.

There are, however, many other frameworks that you can use. Toyota, for example, use a competence level system of:

Level 1: 'I have no knowledge at all';
Level 2: 'I can perform the work only under the detailed instructions of senior engineers';
Level 3: 'I can perform the work under the guidance of an assistant manager or senior engineer';
Level 4: 'I have technical ability and knowledge to a certain degree, and can perform the work by myself';
Level 5: 'I have enough technical ability and knowledge, and have the ability to make improvements'.

Take a little time now to design a competence level system of your own. It does not need to have five levels; it can have as few or as many as you like, but keeping it simple has obvious advantages.

Now return to the key competencies that you decided on for yourself. Assign each one of them the level that you think is appropriate, giving careful thought to the justification for the level which you assign. Think through a number of occasions when you have used that particular key competence. How did your confidence and success in using it relate to the level which you propose to assign?

4.4 Authorising persons

You may well need some form of quality assurance for your key competencies; the need for this will depend on the circumstances.

If you are self employed, for example, you may consider it perfectly adequate to certify yourself. At the other end of the spectrum I have, for example, done consultancy work for a major railway company using competencies as the basis for licensing employees for loss critical jobs; in this case we had to devise a very rigorous system for ensuring that, as far as is reasonable, employees are certified as having a minimum level of competence before they are allowed to perform loss critical tasks.

The more critical the competence, the more important it is to have a rigorous system of ensuring that an individual is correctly assessed. However, the cost of doing this will be correspondingly greater, so a balance needs to be struck.

In my experience, for most professional engineering competencies, it is perfectly adequate to arrange for one or two experts to make a judgement, based on their knowledge of an individual, on that individual's level for any particular competence. This is, in other words, a system of peer review. The software shown in Figure 4.3 shows the name of the person who has agreed with each of my levels of key competence.

There is, however, one other very important factor to take into account. Just because I am judged competent now, does not mean that I will remain competent at that level for ever more. I may raise my level of competence over time through greater practice and experience. Equally, if I do not use the competence, or keep it updated as a result of new technology or circumstances, it may well reduce. It is therefore important that my competencies are regularly reviewed. Again, the frequency with which this needs to be done must be a matter of professional judgement, but the software shown in Figure 4.3 enables variable dates for review to be entered. One # sign appears on the right of the data in the brackets if the current date is within 30 days of the review date, and two # signs appear if the review date has already passed.

One of the great advantages of using software for personal development plans in this way is that the appropriate people (such as the individuals concerned, their managers, mentors, coaches, etc.) can be alerted if review dates are either imminent or passed. Figure 4.4

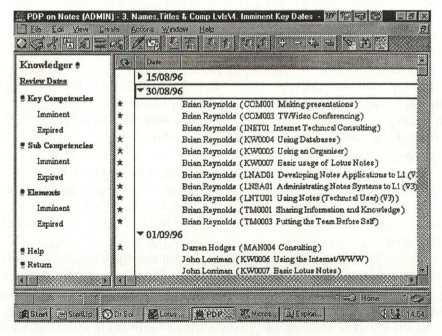

Figure 4.4   An example of lists of competencies identified as about to expire by Knowledge Associates' PDP on Notes software

shows a typical example. Clearly from the point of view of providing quality assurance on the ongoing validity of competencies, software has a vital role to play.

In Figure 4.3 the software shows both the current level of competence (the first figure in the brackets) and the target level of competence (the second figure in the brackets). One of the main purposes of personal development plan software is to enhance the learning process. By targeting the learning to specific improvements in competencies, the learning can be much better focused. As we will also see in Chapter 8, another key role of such software is to link learning experiences directly to competencies and thus to monitor whether they are being maintained, broadened and improved. This requires a partnership between individuals and their managers, as well as any other key players, such as mentors, coaches, the organisation, training providers, etc. Software should only be used as a tool to enable them all to maximise their respective contributions.

4.5 Subcompetencies and elements

The descriptions you will have chosen earlier in this chapter for your key competencies will necessarily have been rather broad. Now is the time to think through the subcompetencies into which each of your key competencies can be divided.

I show an example in Figure 4.5 from my own PDP software of the subcompetencies which I have used for myself for my 'Making presentations' key competence. Again, you will see that it has exactly the same format as the key competence screen shown in Figure 4.3, with subcompetence current levels, target levels, authorising person and review dates.

So now spend a little time looking through each of your key competencies and breaking them down into as many subcompetencies as you think necessary. Assign each of these with a level of competence.

In turn, each subcompetence can be subdivided into a number of elements (of course this process could go on ad infinitum, but a system of key competencies, subcompetencies and elements should be quite sufficient for most purposes). Figure 4.6 shows how my own 'Making presentations—formally to major conferences' sub-competence is in turn divided into a number of elements. Note that you do not necessarily need to have any or many elements within any subcompetence, or to break any key competencies down into

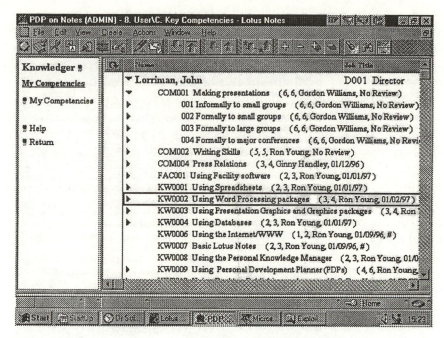

Figure 4.5   An example of some of the author's subcompetencies

subcompetencies. Just develop a system which suits you or your organisation.

Now assign to each of your elements a level of competence.

4.6 The link with appraisals

In my experience most organisations have appraisal systems. Equally, in my experience, almost none of them are very effective. For example, at the US Society for Human Resource Management conference in 1995, Professor Jerry Harvey from George Washington University told delegates that there was 'not a single study' suggesting that appraisals boost performance. 'Those doing it don't want to. Those having it done to them don't want it,' he said. So what has gone wrong?

Most appraisals are based on the management by objectives principles advocated by the American management guru Peter Drucker. He believes, quite correctly, that every employee should have clear targets agreed with their managers against which their performance is reviewed. The meetings at which such targets are reviewed, and new targets set, have become known as appraisal meetings.

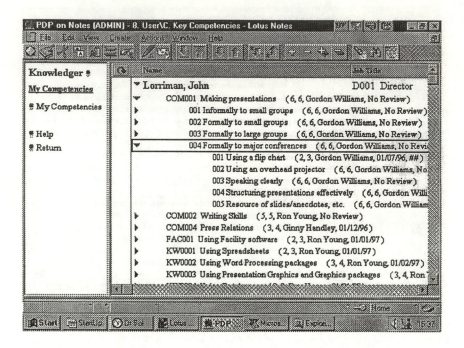

Figure 4.6   Some of the author's subcompetencies divided in turn into elements

In my view the reason that most appraisal systems do not work very well is that they tend to concentrate largely on the targets themselves. I believe that appraisals should actually be about three factors, as shown in Figure 4.7:

- job performance;
- competence development;
- career development.

I draw an analogy with Fleming's Left Hand Rule for motors, as shown in Figure 4.8. Analogies can be drawn between:

- job performance and force;
- competence development and current;
- career development and flux.

The only reason why a motor turns, after all, is because there is an ongoing current flowing within a flux which creates a force, as every school pupil knows. So why should anyone expect an improvement in an individual's job performance unless there are corresponding improvements in the competence development and career development?

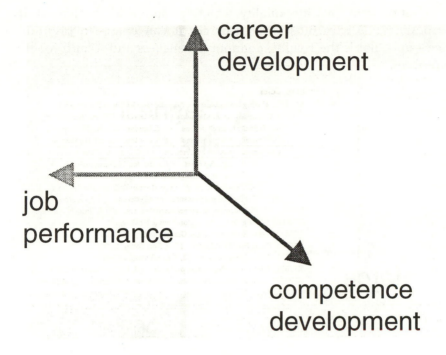

Figure 4.7   Appraisals—the three factors

The only way, I argue, in which appraisals can operate effectively is if they are closely linked to ongoing competence development and career development. Thus, identifying your key competencies, subcompetencies and elements, and those of your staff, is an essential starting point for any appraisal system to work.

Do please note that word, ongoing. It is not enough to identify your competencies at a point in time and leave it at that. You must continually monitor how these competencies are being maintained, improved and broadened. You can use appraisals, personal development plans and coaching and mentoring skills to enable this to happen, and this process will be covered in more detail in future chapters.

4.7 Safety and quality

There seems to me to be a very simple link between an organisation's success and competencies, as shown in Figure 4.9.

The only way in which an organisation is likely to get customer satisfaction is by providing quality in every aspect of its business. If an organisation can provide quality which is superior to that of its competitors, then in most circumstances it can charge a premium price—and that is the road to commercial success and wealth for the employees.

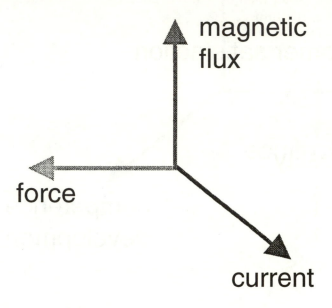

Figure 4.8   Fleming's Left Hand Rule for motors

In order to get the best quality throughout the organisation, every employee must be as competent as possible. Just as importantly, these competencies must be enhanced in every possible way.

An important subset of quality is safety. Safety is a form of quality where a failure is likely to mean that people are injured or even killed. But the principles are identical. Again, emphasis on competencies is essential.

4.8 Job descriptions versus competency descriptions

One of the management tools developed since the 1960s has been the idea of giving each employee a job description. The argument was that each employee should have their area of responsibility clearly defined. Unfortunately, such thinking is rather restrictive. It fails to take into account the range of abilities that each individual has which they do not actually use in their current job, but which might nevertheless be of value to their organisation. Nor does it take into account the need which most people have to develop their competencies gradually over a period of perhaps several years; this is much more difficult to achieve if the emphasis is largely on existing responsibilities and competencies, rather than on those needed by the individual and the organisation in the future.

Customer satisfaction

Quality

Skills

Figure 4.9   The link between success and competencies

One of the benefits from the PDP software shown earlier in this chapter is that it can also be used as a competencies database. Organisations can then move from job descriptions to competency descriptions. They can employ people in a much more flexible way, increasing their job satisfaction, and also do some sensible staff planning, looking at the current stocks of competencies in the organisation and anticipating how these need to change over the next few years.

4.9 National Vocational Qualifications

In the early 1980s the UK found that there was a plethora of examining bodies: City & Guilds, TEC, BEC, RSA, etc. The Government, therefore, set up a Commission of Enquiry, as the British are all too fond of doing when there is a problem, under Sir Oscar de Ville.

When this produced a report in 1986, it in effect concluded that it was politically too difficult to amalgamate these various overlapping examination bodies (which in my view was a serious mistake) and instead suggested that a competence-based system should be introduced. The argument was that both employers and examining bodies should use a framework of competencies and that these should measure in some objective way the actual performance of individuals, rather than concentrating on the results of academic exams.

This was an argument which convinced the Government, and the National Council for Vocational Qualifications (NCVQ) was set up in 1986. During the late 1980s and early 1990s a horrific bureaucracy was spawned—vast numbers of lead bodies to devise sets of competencies for every sector of industry, commerce and government, etc., which anyone could think of, with all too little co-ordination between them.

The language devised by the NCVQ, which introduced National Vocational Qualifications to the unsuspecting British public, was almost designed to confuse: key roles, units, elements, performance criteria, range indicators and underpinning knowledge and under-standing (see Figure 4.10 for an example). I well remember the training director of a highly regarded large industrial company asking me to explain to him in 1991 what NVQs were all about. Even by the end of 1995, a survey of employers by the Institute of Employment Studies showed that more than half of those questioned were not interested in using NVQs.

Amongst the mistakes made by the NCVQ was the failure to recognise the important need to underpin competencies with the necessary knowledge, a mistake which was put right in the early 1990s. However, NVQs are still all too often awarded for life, and I believe that this needs to be changed. In addition, the system assumes that competencies are binary and there is no standard system in place for levels of competence, although again there are signs that the NCVQ is beginning to recognise that this must happen.

Having made these criticisms of NVQs, I nevertheless recognise that the objective of identifying the competence of everyone within a national system is a valid and important one. The Engineering Council, the IEE and most other institutions, as well as most employers, strongly support the NCVQ's objectives and, in any case, NVQs are politically here to stay in the UK. In places such as Australia and Hong Kong, a modified approach to NVQs is developing and many other countries may also adopt this approach.

My advice, therefore, is that you use the framework described earlier in this chapter to suit your own needs. If those needs encourage you to link into NVQs, then almost certainly you will find the NVQ system

Key Role: E51. Evaluate and report on the quality of engineering equipment, systems and services.

Unit ES1.1. Evaluate the design and construction of engineering equipment, systems and services.

Element: ES1.1.1 Verify the compliance of designs and prototypes with specifications.

Performance Criteria:
a. Identification of compliance is made against agreed standards.
b. Identification obtains and draws upon all relevant information and details.
c. Identification of compliance is accurate and makes a balanced judgement over all pertinent factors.
d. Identification is justified with valid, reliable and sufficient supporting evidence.
e. The timing and nature of further investigations needed to resolve difficulties in forming clear judgements of compliance are cost-effective and agreed by all parties.
f. Negotiation and discussions are conducted in a manner that maintains goodwill and co-operation.
g. Decisions are communicated in good time and to the right people for action.
h. Additional specialist information is obtained where necessary.

Range indicators:
Verification relates to: plans; intended manufacturing processes; intended materials; intended quality assurance procedures.
Standards and specifications: national; international; organisational.
Negotiation relates to: standards to be applied; information made available and required; further investigation in the event of difficulties.
Factors affecting judgement of compliance: technical details; timescales; manufacturing/operating/maintenance context.
Designs and prototypes: mechanical; electrical; structural, fluid; dynamic.

Underpinning knowledge and understanding
Principles and methods relating to: design, construction, installation and operation of engineering equipment, systems and services.
Data relating to: standards applied.

Figure 4.10 An example of NVQ-speak—quoted as Appendix 1 of the Engineering Council's 'Competence and commitment' discussion document

sufficiently flexible to give you some form of national certification, if that is what you want. Be prepared, however, for possible battles with bureaucrats who insist that you use their jargon. This will be an excellent test of your diplomacy, patience and self control—so see it positively and try to find some sort of sensible compromise with the NVQ experts; success (or even failure) will in any case provide you with interesting material to put in your personal development plan, as part of the recording of your own self development.

4.10 Summary

Competencies are critical to the success of individuals, organisations and even nations. Japan has long had an excellent and detailed approach to maximising the competencies of individuals in every possible way, and many Western companies, such as Motorola, have achieved outstanding success using the same approach.

Use this chapter to start identifying your own set of competencies and then start thinking about how you would like to see these develop in years to come.

A well regarded book that is worth reading to understand the strategic implications of competencies in terms of an organisation's success or failure in the marketplace is 'Competing for the future—breakthrough strategies for seizing control of your industry and creating the markets of tomorrow' by Gary Hamel and C. K. Prahalad[5].

As Figure 2.1 shows, it is the total context of CPD which matters. Aspects such as the roles of coaches and mentors, as well as the means which an organisation uses to maximise its learning processes, will be covered in more detail in later chapters. At the heart of Figure 2.1, however, is the commitment of each and every individual in an organisation to their self development using personal development plans, and it is this subject which we will look at in some detail in the next chapter.

Chapter 5

Designing and using personal development plans

'We will move through life accumulating portfolios of competencies and intelligences, portfolios which should start, but not finish, at school'—
Professor Charles Handy

By reading this book so far you should by now have accumulated quite a number of ideas and documents, such as the competencies set which you identified in the last chapter.

Lay all these out on a table top and see if there is any overall theme. Ask yourself which seem important, and might require immediate action, as well as which are really better classified as longer-term strategies and dreams.

In this chapter we will look at different approaches to designing personal development plans. You can then decide, depending on whether you are looking at this subject as an individual, a company or a professional institution, on the benefits of either using an existing design or designing your own based on the ideas described.

5.1 Me-charts

Two of the key lessons that I have learnt over the years are that:

1 Learning is much more effective if it is fun.
2 Learning using some form of graphics is much more likely to be both meaningful and memorable.

One of the ways, therefore, in which I often open up the workshops that I run on the subject of CPD is by asking the delegates to take a flip chart each and draw a me-chart. This is a pictorial representation of you as a complete person and is quite a good way to start—by thinking through the elements of your life that really matter to you. You need to see yourself in the round. For example:

- what is the range of your activities at work?
- what are your hobbies?
- what are your values and most important beliefs?
- who are the immediate members of your family and how do they see themselves as supporting your CPD?

Figure 5.1   The author's me-chart

I attach my own me-chart as Figure 5.1. In the centre is a heart, inscribed 'IEE', since so many of my activities and so much of my learning seems to centre around the Institution. Apart from this book for the IEE publishing arm, my IEE activities have led over the years to strong links with The Engineering Council. I have responsibility for the IEE's links between the centre in development in Belarus and the South East Midland Centre, and have chaired both the South Midlands and South East Midland Centres of the Institution. Looking at my me-chart clockwise from the bottom right, the book shows that this is my third major publication. The two match people represent my wife Jill and myself partnering each other through life. The map of the world shows my commitment to and strong links with the International Association for Continuing Engineering Education (IACEE), of which I am a Council and Executive Committee member, and where I chair the professional development working group. Around the world my main interests centre at present on the UK and Turkey, where I do a lot of consultancy, on Hong Kong, which I see as the gateway to China and on Japan. At the top of the me-chart is a computer showing my commitment to developing software versions of PDPs based on DataEase for Windows and Lotus Notes software.

So that is me in a nutshell. The chart lets me sit back and decide whether this is really where I want to be at this stage of my career (it is!) and think through which parts of my CPD should be given the greatest priority in the future, both in the short term and in the long term.

Now draw your own me-chart and share it with someone else— maybe a friend, your partner, your manager, a colleague, a coach or a mentor. By discussing it with that person, and thinking about their reactions and questions, you will be that much better placed to start devising a structured document which you can call your personal development plan.

5.2 What are personal development plans?

As Professor Charles Handy says in the quote at the beginning of this chapter, we will all increasingly need to have some form of portfolio of our learning, achievements and career aims. Many pupils have left British schools for a number of years now with documents called National Records of Achievement, which are just such portfolios developed by the pupils to show potential employers their

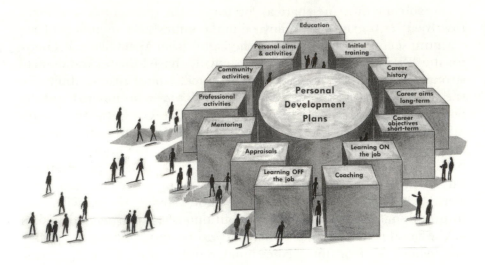

Figure 5.2   Some examples of the possible contents of your personal development plan

achievements in the round—not only their academic achievements, but also all their other interests, successes and objectives in life.

Such a portfolio makes a lot of sense, not only when leaving school but throughout our lives. The names of such portfolios vary; consultancy work I did for Abbey National led to the introduction in that organisation of what they called a personal development file, while the one I helped the IEE to develop is called the professional development record. Call it whatever you or your organisation wishes, but my own preference is for the term personal development plan, or PDP for short.

Figure 5.3   Some different examples of personal development plans

To me the first word, personal, is very important because this shows very clearly where the ownership lies—with the individuals themselves.

Figure 5.2, reproduced from The Engineering Council's 'CPD—the practical guide to good practice', shows some of the wide range of contents that you might wish to include in your own PDP. The style, size and format of a PDP can vary enormously, with some examples shown in Figure 5.3, reproduced from the same Engineering Council publication.

5.3 The story of the PDR and the career manager—two of the first national PDPs

In June 1986 The Engineering Council published a report entitled 'A call to action—continuing education and training for engineers and technicians'. At that time what is now generally called CPD was usually referred to as continuing education and training (CET). As the foreword to the report from Sir Francis Tombs, Chairman of The Engineering Council, said:

> 'This report is about action to achieve a radical change in attitude towards improved CET in British industry and commerce by employers and individual engineers and technicians. It is also about action by others with a major role to play: by the professional institutions, by providers of CET, by the Government, by training boards and associations, by trade unions and, of course, by The Engineering Council. We want less exhortation and more achievement. This booklet shows how you can help.'

One of the suggestions for achieving these aims was that:

> 'The Engineering Council wishes to encourage the recording of continuing education and training by all engineers and technicians in personal record log books.'[6]

At the time I was a member of the IEE's career development committee and I suggested that we should develop such a document for IEE members.

The document that I designed together with members of the IEE staff was named the 'Professional development record' (PDR). It is shown in Figure 5.4 and has ten sections covering:

- education;
- employment;
- career objectives (long term);
- career aims (short to medium term);

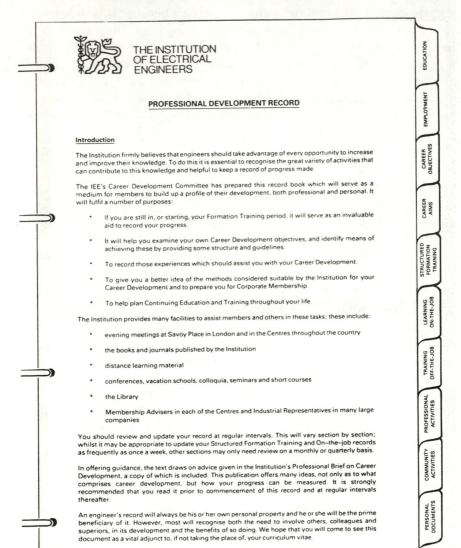

Figure 5.4   The IEE's PDR

- structured formation training;
- learning on the job;
- training off the job;
- professional activities;
- community activities;
- personal development.

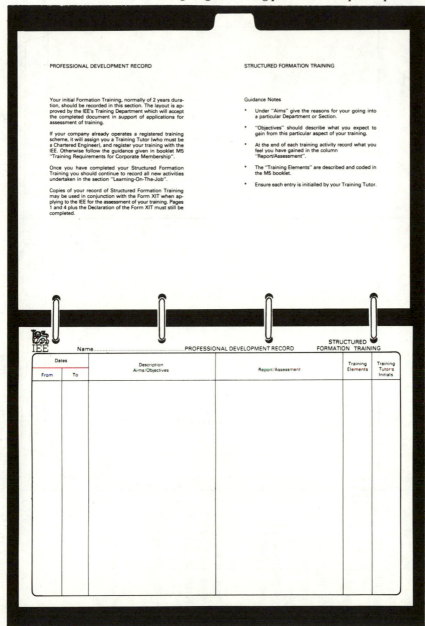

PROFESSIONAL DEVELOPMENT RECORD

STRUCTURED FORMATION TRAINING

Your initial Formation Training, normally of 2 years duration, should be recorded in this section. The layout is approved by the IEE's Training Department which will accept the completed document in support of applications for assessment of training.

If your company already operates a registered training scheme, it will assign you a Training Tutor (who must be a Chartered Engineer), and register your training with the IEE. Otherwise follow the guidance given in booklet M5 "Training Requirements for Corporate Membership".

Once you have completed your Structured Formation Training you should continue to record all new activities undertaken in the section "Learning-On-The-Job".

Copies of your record of Structured Formation Training may be used in conjunction with the Form XIT when applying to the IEE for the assessment of your training. Pages 1 and 4 plus the Declaration of the Form XIT must still be completed.

Guidance Notes

* Under "Aims" give the reasons for your going into a particular Department or Section.

* "Objectives" should describe what you expect to gain from this particular aspect of your training.

* At the end of each training activity record what you feel you have gained in the column "Report/Assessment".

* The "Training Elements" are described and coded in the M5 booklet.

* Ensure each entry is initialled by your Training Tutor.

IEE   Name................................   PROFESSIONAL DEVELOPMENT RECORD   STRUCTURED FORMATION TRAINING

Dates		Description Aims/Objectives	Report/Assessment	Training Elements	Training Tutor's Initials
From	To				

Figure 5.5   A sample section of the IEE's PDR

Figure 5.5 shows an example of a section from the PDR. You will see that the design was deliberately extremely simple; brief instructions, together with some simple guidance notes on the divider opposite each section, and a very simple design to each page.

Area 1. YOUR PRESENT JOB

The aim of this part is to identify where action is needed to maintain and improve your contribution to present job performance.

On the opposite page you should analyse your job under three headings:

—What are the main areas of work of your present job?

—For each area, what is the scope for improvements to ensure you carry out your present job effectively?

—What improvements are required (in terms of additional knowledge and skills needed)?

Main areas of work:

This is what you aim to achieve at work: it is why your job exists and how you bring added value to your employer. Refer to your job description if you have one.

You may find it helpful to break down your job into three areas:

  (a) Technical/ Scientific

  (b) Personal effectiveness/Communications

  (c) Commercial/Business

Area 1. Present Job		
Main areas	Scope for improvements High/Medium/Low	Additional knowledge and skills needed

Figure 5.6   An example section from the Engineering Council's 'Career manager'

In designing the PDR we looked at a number of such documents which were being used, including one used by the Institution of Mechanical Engineers. We also kept in close touch with The Engineering Council and the then senior executive, CPD, Bernard Dawkins, who was designing and piloting a career management document which they called 'The career manager'.

'The career manager' was very much designed to structure one's career and contained the following sections:

Section 1: Identifying career development needs
Area 1: Your present job
Area 2: Future roles
Area 3: Professional and personal development
Key development aims

Section 2: Career action planning
Activity 1: Proposed development actions
Activity 2: Employer review
Activity 3: Advice/counselling
Activity 4: Career action plan

Section 3: Recording achievement
Part 1: Record of CET development activities
Part 2: Career record

An example section from 'The career manager' is shown in Figure 5.6.

In contrast, the IEE's PDR was designed to be rather broader. Its purpose is to act as a record of learning right from the start of an engineer's career and integrates, for example, with the Institution's documentation for recording accredited training, with space for mentors to put their initials and comments at the end of each period of formalised training. I have always argued that IEE members should continue using a PDR throughout their careers, and maybe even after they retire; jokingly I have suggested some might even want theirs buried with them!

The PDR was piloted by some 200 volunteers at GEC Telecommunications (where I was controller of training at the time) and by a rather smaller number at the IBA and the Electricity Council. After it had been in use for some months I sent out a questionnaire to the GEC Telecommunications participants. The comments included:

'Listing areas of training and experience gives a positive feeling of achievement.'

'It has allowed me to see at a glance what jobs and training I have been involved with, which puts the work in perspective. It helps in considering future career development.'

'I have positively targeted some career goals. Collation of papers and relevant material into one comprehensive file has been useful.'

I was also involved in helping Chris Senior, Bernard Dawkins' successor at The Engineering Council, to pilot 'The career manager' at GPT; this was a small part of a pilot which used a total of 63 organisations involving over 1000 individual engineers and technicians. The report on this pilot was published by The Engineering Council in March 1991[7]; some of the comments by individuals using it were:

'Allows me to plan my career in a consultative way.'

'Acted as a catalyst for ideas and action.'

'Sets your sights further into the future.'

In addition, the report said that:

'Over 80 % of participants found the document was useful/very useful in identifying and planning their CET.'

'The career manager' has continued to be used by quite a number of organisations since.

In contrast, the IEE's PDR was launched commercially in October 1987 at a price of around £15. By December 1990, 5000 had been sold, with the ten thousandth sale being achieved in December 1992 and sales passing 15 000 in 1995. In 1997 a Java-based PDR on Disk was launched, developed by Knowledge Associates.

For several years the Institution of Electronics and Electrical Incorporated Engineers (IEEIE) used the PDR under licence from the IEE, until in 1992 the organisation launched its own 'Record of professional development' (RPD) which it sold to members for a similar price to the IEE's PDR, and a free copy of which was given to all new students joining. The RPD was then made available to the Institution's 27 000 members free of charge in the form of a computer disk compatible with all Windows 3.x software, to which there was a very positive response. A sample screen is shown in Figure 5.7. Further details can be obtained from the IEEIE's CPD department via e-mail at: ieeie@dial.pipex.com.

In 1993 the Institution of Mechanical Engineers replaced its own personal planner with the IEE's PDR, and in 1995 the Institution of Biological Engineers started using the PDR under licence. Several companies, including my own GPT at the time, tailored the PDR to their own requirements in agreement with the IEE.

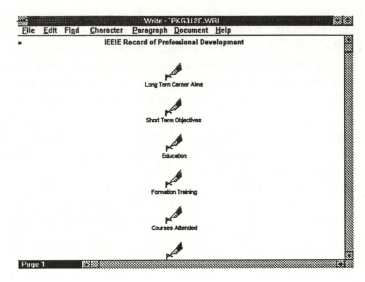

Figure 5.7   *A sample screen from the Institution of Electronics and Electrical Incorporated Engineers' 'Record of professional development' software*

5.4 Starting to design your own PDP

You now have sufficient background and ideas, I hope, to start designing your own PDP.

If you are reading this chapter as an individual, then my advice is to start with something simple, perhaps designing a few master sheets in a loose-leaf folder. You will find out whether or not it is a good design for you very simply—by using it. Alternatively you might decide to spend a little money and buy a proprietary PDP, such as the IEE's PDR or the PDP software described in the previous chapter. If you are a member of the IEE or IMechE you may wish to use the PDP on the Web software described in Chapter 8 (page 135). Either way, discipline yourself to put in entries as regularly as possible—at least once a week and ideally once a day.

On the other hand, you may be reading this chapter from an organisational point of view and wish to develop a corporate approach to PDPs. In that case, my strong advice is that you need to involve the participants as much as possible at every stage in the decision to use PDPs, the design, the context, the launch, etc. The one failure I had in this area was when I had such success with the IEE's PDR that I persuaded my personnel colleagues in GPT to issue copies to all 200 to 300 technicians and graduates joining the company one year. Even though they were all briefed on the document, many never even started using it; the British simply do not like being told what to do. Very significantly, those people who did use their PDR regularly were generally the ones whose managers showed both interest and support—which is where my concept for Figure 2.1 originated.

In contrast, in a number of organisations where I have worked since as a consultant, by using a group of enthusiasts to help me to design and launch a PDP designed to meet the specific needs of their organisation, the document has spread far more quickly than anyone expected. In one organisation, for example, 1000 copies of the new PDP were printed for a pilot; the pilot never took place because word of the new document was spread by the internal enthusiasts who had designed it, and managers were soon telephoning the management development manager asking for copies. Stock was exhausted before a controlled pilot could be done. In another organisation, it was originally planned that one department would pilot the new PDP for three months, another three would follow on in the next three months and the other 20 departments thereafter; in practice many people from these other 20 departments were soon demanding to be involved as soon as possible and the pilot thus turned into a full-scale launch, by popular demand, very quickly.

Never forget that the context in which PDPs are used is vitally important to their success. Remember the three factors in Figure 2.1:

● individuals with a missionary zeal towards their self development, using PDPs;
● managers who see their most important role as coaching and developing their staff;
● organisations which both reward managers who do develop their staff effectively and which ensure that every opportunity is taken to maximise learning within the organisation.

One set of questions that individuals might like to include in their PDPs and ask themselves after every learning event is:

● what did I do?
● what did I learn?
● how might I apply this, now or in the future?

When you are considering a job move, you might include the following three questions in your PDP analysis of whether or not it is the right career step for you:

● is it going to be interesting?
● are they people I can work with?
● am I going to enjoy myself?

As an organisation, you might design corporate PDPs in such a way that they encourage teams or groups of employees to ask these three questions:

- what has gone well from which we can learn?
- what has not gone well which we can improve in future?
- what issues, ideas and concerns do we have?

In other words, see the PDP as acting as a focus for both individual and corporate learning. By learning faster than the opposition, you will not only survive, but may well become number one in your field in the world.

5.5 How widely are PDPs used—and in what context?

The 1992 Engineering Council survey showed that nearly three quarters of engineers undertook some form of CPD activity.

However, despite an increase in the level of CPD undertaken during the previous three years, a huge majority, well over eighty per cent, felt that there was room for improvement in their own professional development schemes. Another key concern was that only one in four survey participants had plans for improving their own CPD.

In 1993, a survey of 4500 members of the Institution of Civil Engineers concerning CPD produced a 36.7 % response, demonstrating the interest in this topic. The results showed general support for CPD and in particular that:

- more should be done by employers to provide CPD opportunities;
- the Institution of Civil Engineers should do more to advertise and promote CPD;
- more information is sought as to what constitutes CPD;
- urgent action is needed to provide opportunities for distance learning;
- the recession and consequential shortages of time and money have been the principal causes of the reduction in CPD activity;
- action should be taken to help members to keep themselves up to date;
- special attention should be paid to the needs of specific groups, including overseas members, women returning to work, the unemployed, older members and associate members.

Some of the key findings were that:

- only twenty per cent had a written personal development plan of any description;
- under thirty per cent were given any sort of formal company performance appraisal;
- only thirty five per cent had access to a formal organisation system for the identification of training needs;
- just fifty two per cent of UK based members and thirty eight per cent of overseas members had undertaken any formal training during the past 12 months.

Two comments from the respondents were:

> 'It is good to see a policy on CPD being developed for ongoing training. Keep up the good work.'

> 'CPD is essential to the development of our profession and should be actively encouraged by the Institution.'

Alan Davis and Mike Fox, regional CPD managers for The Engineering Council in the South West at the time, carried out their own regional survey of the attitude to CPD among Engineering Council registrants in 1993. A random sample of 6.5 % of the registrants in the region were sent a questionnaire, of which thirty four per cent were returned.

Some of the key results were:

- younger registrants formulate development plans to a greater extent than older ones, mostly via employer initiated schemes, and overall an encouraging seventy four per cent were identifying their learning needs;
- CPD activities were reported as taking over 160 hours per person per annum;
- about sixty per cent of registrants up to the age of 54 years said that they were willing to increase their annual investment in their CPD by 100 hours per annum;
- over seventy per cent of the sample reported positively on the value of their CPD activities to their companies;
- over sixty per cent said that they were keeping a formal record of their development activities.

A 1993 survey of 500 corporate members (of whom 174 replied) of the Institute of Highway Incorporated Engineers (IHIE) showed:

- half of the respondents had some form of personal development plan forming part of an organised scheme;

- as many as forty two per cent of the under-30s had no plan—not even 'in their heads';
- one-third had access to a formal appraisal-type system and had discussed CPD with their employer during the last 12 months;
- half of the members had undertaken no formal training in the last 12 months;
- nearly two-thirds said that there were areas where they had been unable to increase their competence in spite of a desire to do so; the reasons included:

cost (employer)   28 %
time (personal)   20 %
no suitable opportunities   21 %
no personal motivation   5 %

- computers and management were the main subjects named as requiring help;
- half of the respondents did not record their CPD—not even in a running CV, although all had been sent an IHIE personal record.

Overall there was an overwhelming desire by members to undertake CPD but this was thwarted by a lack of interest by employers, lack of funds to finance training and insufficient time.

The context in which PDPs are implemented is therefore vitally important.

5.6 Competence and commitment

In January 1995 The Engineering Council published a policy statement document, presenting proposals for a new system of engineering formation and registration, and entitled 'Competence and commitment'[8]. Among its proposals were that registration should require engineers and technicians to demonstrate and maintain both:

- the competence to perform their professional work to the necessary standards;
- the commitment to maintain that competence, to work within professional codes, and to participate actively in the profession.

It is proposed that competence will be developed in three parts: foundation learning, specialist learning and competence in employment against the background of clearly defined occupational standards.

To be assured of commitment, The Engineering Council will require that candidates seeking registration maintain, and submit for audit on request:

- action plans and log books, showing how a profile of competence is being maintained and developed, as technology and work roles change;
- portfolios of work showing how the codes of conduct and the risk and environment codes have been applied and conflicts resolved;
- contact with professional colleagues in learned society activities, mentoring of new entrants and other professional work for the public good.

To enhance the profile of UK-based European engineers, it is proposed that candidates for the Eur Ing title will, at some time in the future, be required to fulfil the FEANI expectation of fluency in a European language other than the mother tongue. The normal requirement will be language competence to NVQ Level 3, based on the language lead body standards.

While it will take time to debate and action, where agreed, these proposals, it is clear that the engineering profession is moving inexorably towards requiring that all technicians and professional engineers maintain some form of PDP which acts as the focus for their CPD.

Let us therefore now look at an example of a company which has successfully implemented PDPs, taking note of the context in which it did so.

5.6.1 Short Brothers focus on the future

Short Brothers plc, the Belfast based aerospace company, launched a major new initiative—Focus 2000—in 1993 in which there were three key elements:

- value in people;
- customer focus;
- best cost producer.

In the value in people element, continuing professional development was seen as a major contributor in increasing business performance through the process of continuous learning. All of the company's employees were encouraged to maintain a personal development file which would enable them to keep an up to date record of their learning achievements and to monitor their individual personal development plan.

Training committees were established with a key purpose of co-ordinating the identification, design and implementation of the training requirements of each business team which were then directly linked to key company business objectives. Each employee was then encouraged to take ownership of his own development and to participate in regular progress reviews.

Next, let us look at CPD and the use of a PDP, from an individual's view-point, again taking careful note of the context in which it is taking place.

5.6.2 Thames Water Utilities—an individual's viewpoint

Terry Sweeney, a Project Design Engineer with Thames Water Utilities, describes his use of a PDP within the context in his company as follows:

'My department is increasingly involved in producing in-house designs of tender specifications, which were previously subcontracted. As part of the ongoing CPD provided to respond to this and other challenges, the company organises a series of in-house seminars run by our own senior staff. These take place for one day every two months on average, with in excess of ten subject specialists at each.

'I attended a career development course run by consultants eight months ago. We looked at the organisation's values, examined what we would like to do in the future and prepared an action plan. I also read a lot—for example the 'IEE Review', the business section of 'The Times' etc. On average I estimate that I spend about two hours each week on my self development.

'The structure of the company's CPD continuous improvement log emphasises the need to keep up to date with current—and possible future—developments in one's own field. It acts as an effective replacement and further development guide for me since qualifying—which in itself required me to keep a log of my learning. And that's the key—continuous learning and improvement, not just formal training.'

5.6.3 PDPs in a safety context

One important context in which PDPs can be valuable is in the area of safety. Here, therefore, is an excellent example.

The Institution of Railway Signal Engineers launched its new licensing scheme on 25 January 1994, with Sir Anthony Hidden as the guest speaker. (Sir Anthony chaired the investigation into the Clapham Junction Railway Accident in December 1988.) The scheme introduced a system for the licensing of competencies for safety-critical and safety-related tasks in the railway industry.

In his speech at the launch, Sir Anthony said:

'Today is a day to look forward. Anything which has the effect of improving the standard of expertise in, and management of, railway signalling must be applauded as a good thing, and thus I warmly welcome the inauguration of this licensing system.'

Participants in the scheme are required to maintain a continuing professional development folder and licensing scheme logbook as proof of their continuing employment on licensable work. This record of the competent use of skills will be reviewed when licences are due for renewal, and will also assist railway industry employers to satisfy the work authorisation requirements contained in the Railway (Safety Critical Work) Regulations, which came into force on 1 April 1994.

Already 30 activities in design, installation, testing and maintenance have been identified for licensing in signalling and telecommunications. Employers of licensable staff are now able to register with the scheme, and they will receive copies of competence assessment checklists to use in the workplace assessment of their employees, prior to formal assessment by an IRSE approved assessing agent.

The CPD folder has an important role to play in both ensuring that these licensed competencies are maintained and in planning the licensee's career development. In addition, the document provides employers with a verified record of a prospective employee's capabilities and achievements.

This is an example of an institution using a PDP primarily for the purpose of improving public safety. However, let us look at five institutions using PDPs for broader CPD purposes.

5.6.4 Professional development for physicists

The Institute of Physics (IOP) is now operating a professional development scheme, consisting of a two year training period followed by a further two years for consolidation during which experience is acquired. The objective is to increase and enhance the business and organisational skills of the participant, in addition to developing the relevant technical knowledge required by the job.

For every participant, a mentor is appointed who will provide advice and guidance. The mentor is a representative of the Institute of Physics and is, ideally, an employee within the same organisation as the participant. It is the mentor's role to assess the participant's performance against the professional requirements of the Institute.

Initially the participant and mentor construct a career development plan, with the agreement of the employer, which is approved by the IOP. This plan identifies the broad objectives to be achieved to meet the employer's training and development needs, as well as the Institute's requirements for professional qualifications.

Each participant is issued with a professional development record (PDR) book. This remains the property of the participant and contains personal details, academic qualifications and achievements and, as the scheme progresses, a validated record of the training and experience acquired. This book also forms the basis of CPD as the participant's career develops.

The final assessment, which leads to the award of CPhys and full membership of the Institute, is based on the mentor's report, examination of the PDR and an interview. Some participants will also be eligible for chartered engineer (CEng) status.

5.6.5 Foundrymen launch CPD record

The Institute of British Foundrymen launched a new CPD Record Book in 1994 which is being issued to its members. This is part of a policy of encouraging members to undertake a minimum of 35 hours of CPD per year.

Members are encouraged to have their CPD record books authenticated and to return them with their annual subscriptions to the Institute. These will then be validated by the Institute, which will issue certificates to those meeting the minimum CPD requirement.

5.6.6 CPD commitment in Hong Kong

The Hong Kong Institution of Engineers (HKIE) introduced a new record of continuing professional development in 1994, as part of a policy requiring a minimum of 45 hours a year CPD up to the stage of full membership. Thereafter, the individual's CPD record is taken into account when considering any application for fellowship.

The format of CPD activities can include, among other possibilities:

- courses, lectures, seminars and workshops;
- industrial attachments and visits.

Individuals use the new record of CPD document to record these activities, which need to be endorsed by an engineering supervisor or a responsible person.

5.6.7 Plumbing for CPD

The Institute of Plumbing provides all members with a small piece of plastic on which they commit themselves to:

- perform professionally, competently and responsibly;
- safeguard the public interest in matters of health and safety;
- comply with all relevant laws, regulations and standards;
- broaden, improve and maintain their skills, knowledge and personal qualities;
- uphold the dignity, standing and reputation of the Institute of Plumbing and the plumbing mechanical engineering services industry.

To assist members with meeting their CPD commitment, a new system involving authorised CPD providers (AP) has being introduced. These may be individual lecturers/trainers who will present their subject at venues to suit a particular audience. Alternatively, a college or company with an acceptable training centre and staff can be approved as an AP. In each case, course content is assessed and all details entered in an Institute of Plumbing CPD directory.

CPD Record Cards have also been issued to all Institute members and the Institute hopes that many members will exceed the 20 hour CPD recommended minimum per annum. Fellows and members who are registered as IEng or EngTech with The Engineering Council have an additional responsibility in maintaining a record of their progress.

One aspect in which plumbers have every incentive to engage in CPD is where gas is involved; here statutory regulation applies. As from 1 October 1995 all new CORGI (The Council for Registered Gas Installers) approved plumbers have had to be qualified according to the relevant criteria. As from 1 April 1998 all existing operatives will have to meet the same requirements. Approved code of practice courses dealing with standards of training in safe gas installation have consequently been in great demand.

5.6.8 Putting energy into CPD

The Institute of Energy published its policy statement on CPD in 1995 and has produced a career management planner. This planner assists in:

- identifying career needs;
- planning future development;
- recording and monitoring progress.

As the Institute is a small one with around 5000 members and a staff of eight, much of the work in preparing the statement and producing the planner fell to the Institute's professional development group. In carrying out its work it had to bear in mind that its recommendations should be clear, simple and require the minimum of administration in terms of time and money.

The group took very much into account the policy statements of other institutions and The Engineering Council, and its planner is largely based on that originally developed by The Engineering Council.

The Institute of Energy has decided on 35 hours a year as a reasonable minimum attendance on CPD activities. Although it has decided not to make CPD compulsory, members applying for transfer to higher grades will be expected to show evidence of participation in CPD activities. The Institute strongly supports CPD in principle and will continue to encourage members to develop the right attitude of mind.

5.7 Some further ideas

In this chapter we have looked at various ways and various contexts in which PDPs have been designed, used and put into an effective context. We have looked at PDPs from an individual, organisational and institution point of view.

So here are a few more points for you to think about:

- are you one of the many young engineers that I frequently meet who have found that their companies have forced them into a specialisation? What do you really want from your future career? How can you change the situation to achieve your ambitions? How can you use your PDP to analyse the possibilities? Can you, for example, use some of the ideas in Chapter 7 about coaching upwards to persuade your manager to give you the broader experience which you need? Or should you be looking for another job?
- how good are you at reading, writing, understanding and speaking foreign languages? Should this be an important part of your PDP self analysis? It is certainly likely to become an important criteria for achieving Eur Ing status and, in any case, it makes good business sense for companies to have a wide range of employees fluent in other languages if they have international ambitions;
- how much networking do you do? Zeneca is an example of a highly successful organisation which encourages every graduate to network as much as possible right from day one. The company even goes to great lengths to produce a detailed guide to the expertise across the

company, including contacts with special knowledge and areas of knowhow and expertise, supported by the telephone numbers and sites of everyone concerned. You might think of including a section on the network of contacts which you have established as part of, your PDP, and adding to this over time;

- do some cash planning. Many well-known politicians, such as Michael Heseltine, made it a key part of their life plan to accumulate a cash mountain early in their careers, which then allowed them to enter politics with no worries about the financial consequences. If, as it has been for me, your ambition is to move towards setting up your own company, then you will need to plan very carefully how you will put together the funding required. Build these plans, together with careful monitoring of the way in which your wealth is developing, into your PDP;

- plan for the unexpected. When Barings, the UK's oldest merchant bank, went belly up on 26 February 1995 as a result of massive losses in derivatives in its Singapore office, that day's copy of 'The Sunday Times' ironically carried a job advertisement for deputy head of internal audit—a bit late perhaps! How prepared are you for the totally unexpected? This might range from a new boss pointing you out of the company to a headhunter telephoning you about a prestige appointment. The best way to be prepared is by keeping your curriculum vitae up to date, by constantly monitoring the development of your competencies and by a regular review of your long-term career objectives;

- do not underestimate the importance of your interests away from work. In one organisation in which I have introduced PDPs on a consultancy basis, I included a section on 'the balance in my life'. As often as not, even senior managers turned to this section first when they were initially introduced to the PDP;

- what could you do to encourage the next generation to follow you into engineering? Could you help by giving talks in schools or becoming a Neighbourhood Engineer? In particular, if you are female, could you do more to persuade female school pupils to consider engineering as a career?

- could you become more involved in community activities, for example by volunteering to become a school governor or entering local politics? Nissan in the UK was experimenting in 1995 with providing community work for its employees, starting with a pilot group of 200, encouraging them to work in external organisations including charities and the civil service; the idea was to tackle the

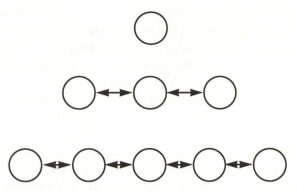

Figure 5.8   The need for horizontal experience and promotion in flatter structures

problem of blockages to upward career movement created by delayering and to explore ways of creating more employment;

- organisations are increasingly delayering and moving to flatter structures, looking to employ staff who are multifunctional. Give careful consideration, therefore, to the need for horizontal promotion—see Figure 5.8. Too many ambitious people suffer a rude, and often terminal, shock to their career when they are promoted vertically too fast. You will need to develop a wide range of competencies at the initial stages of your career if you are to be able to make the best decisions, either managerially or technically, later in your career. Actively plan and seek horizontal broadening of your career; this may be by moving to a new job at the same level or it might be by experiencing a wide range of activities and functions within your existing job;

- organisations would do well to move away from job descriptions, which are restrictive and short term in their inherent concept, and towards competency descriptions for all employees. This will give much greater flexibility, allowing everyone to contribute to the organisation's objectives, and also provide greatly enhanced job satisfaction. In addition, it is likely to provide ongoing broadening experience—or horizontal promotion—for everyone. PDPs then become a vitally important way of maximising the development of all employees within such a fluid and sophisticated organisation.

- if you intend to have a career break to have a family, then use your PDP to plan how you will keep up to date during the time you are away from work. Will you want to go back to full-time work, want part-time work, wish to job share, work from home, or set up your own business, etc?

- recognise that of the 2.7 million companies in the UK, only 9000—a third of one per cent—employ more than 200 people. By contrast, 96% of companies have fewer than 20 staff, and two-thirds of all businesses consist of just one or two people. Indeed, between 1989 and 1991 businesses with fewer than 20 employees created nearly half a million new jobs, while the larger firms saw their workforces decline. Therefore, see your career possibilities in as wide a context as possible, not necessarily restricting your experience to the larger companies.

- organisations are increasingly moving away from employing all the expertise which they need. For a large number of reasons, such as the costs of overheads and flexibility, there is a widespread move towards virtual organisations—these are ones where most of the work is subcontracted out to experts, sometimes on a long-term basis and sometimes on an interim or temporary basis. This is happening in every area of organisations, including in research and development. (The OECD basic science and technology statistics show that the ratio of external to internal funding of research and development has been increasing by two to three per cent each year in the 1990s.) Professor Charles Handy has predicted that by the start of the next century less than half of the industrial world's workforce will be in full-time jobs. In addition, there are now more and more people, myself included, who much prefer the freedom of working on their own terms with a variety of organisations. Those who do will then find, I am sure, that they will need portfolios of areas of expertise to show potential clients what they are able to offer; this is likely to be in the form of a curriculum vitae, areas of expertise (a competence set), previous clients for whom you have worked, etc. The easiest way in which you will be able to do this, I believe, is by using a PDP in the form of software which can easily be electronic mailed or faxed to potential clients. Some of the options have been described in the previous chapter and more will be said about these exciting possibilities in Chapter 8;

- the use of software for PDPs is certainly something which I strongly recommend. You can, for example:
  —link learning experiences directly to your competence set;
  —monitor your learning to ensure that there is a balance between your short-term and long-term, and technical and nontechnical learning;
  —use the powerful reporting facilities to monitor how much time you are spending on learning, whether you are obtaining learning linked to all of your key competencies, and so on;
  —use the software as a skills database in an organisation.

5.8 Keep your CV up to date

You never know when your curriculum vitae (CV) may turn out to be useful, so keep an updated version as part of your PDP. In many cases people can now expect to have eight or more employers during their working lives, so be prepared for the unexpected—either from a positive or negative point of view.

There are many different possible formats, but the one I prefer and have used successfully on many occasions has the following structure:

- a front sheet with my name, qualifications, age, marital status, address, telephone number and fax number. These basic details are then followed, for the rest of the front sheet, by a one sentence summary of my background and a number of bullet points giving my special strengths;
- a second page with basic details of my career in reverse order, particularly highlighting my successes in each job;
- a third page with details of my education, fellowships, etc., professional activities, publications and personal interests and activities.

Keep your CV to the shortest length possible. No potential employer likes to wade through pages of details. The only purpose of a CV is to hook the potential employer into wanting to meet you, and you will do that most successfully if you produce a short punchy CV, backed up by a tailored one page covering letter explaining why you are a highly suitable and well-qualified candidate for the job.

5.9 And if redundancy should hit you...

More and more professional engineers are unfortunately experiencing the pain of forced separation from their employers.

First, do not panic. If you have followed the principles outlined in this book, then you are in an eminently good position to take full advantage of the opportunities now open to you. Think carefully about what you would really like to do next in your career. Write these thoughts down in your PDP and discuss them with other people, probably starting with your family.

If you decide to apply for another full-time job, then do not just apply for advertised vacancies in the national and professional press; also use the network of contacts that you will have identified in your PDP. Ask this network of contacts for any ideas on job opportunities in the area you are looking for, rather than embarrassing them by directly asking for a job.

Then try to speak to the right person in the organisation to which you wish to apply. Use your people contact skills to establish a rapport over the telephone with the secretary of the person that you wish to meet. Remember, you are using and developing your marketing skills to sell yourself—and few things can surely be more important than that!

The IEE provides a job location service with the aim of assisting unemployed members in the UK to locate suitable employment. The IEE also provides seminars on the subject of seeking a new job. Contact the career development manager of your own institution to see what advice and support might be provided.

Think about whether you are prepared to offer your skills on a temporary basis, thus reducing the commitment and risk for your potential employer until you are better known.

Tailor your CV and covering letter very carefully, taking several hours if necessary to adapt both to the specific requirements of the job for which you are applying.

Above all, see all this as a challenge which is an important part of your CPD. Believe in yourself and let everyone else know that you do too. And avoid any feelings of bitterness; think positively and recognise that you are, sadly, simply in a position in which increasing numbers of other people are inevitably also finding themselves.

Remember that redundancy is now so common that a lot of the managers who might interview you for a possible position with their company may well have been through the same experience themselves!

5.10 Summary

It is very much in the interests of both individuals and organisations that a high priority is given to CPD. A survey by the Institute of Management in March 1995[9] showed that, although the majority of managers feel that they are on top of their daily tasks, many fail to devote sufficient time to long-term strategic planning and their own personal development needs. As the report noted, this short-term emphasis is potentially damaging both to UK competitiveness and to the careers of individuals.

The best possible way, and indeed I argue the only effective way, to plan and structure your CPD is with the use of a personal development plan. I hope that by reading this chapter you and your organisation are now well on the way to feeling comfortable with using such a document widely and regularly.

But, in addition, as this book has constantly emphasised, the context is vitally important to successful CPD, including the use of PDPs. Perhaps the most important context of all is the role of managers as coaches, as well as the use of mentors, and this will be the topic of Chapter 7.

Chapter 6
Creating the right learning opportunities

'Total quality begins and ends with training'—Professor Kaoru Ishikawa

I confess to a great admiration for the Japanese approach to training. The emphasis in Japanese companies is very much on learning on the job, with managers predominantly performing the role of coaches. In contrast, the American approach tends to be one of throwing money at the issue, using expensive off the job training, with managers encouraged to behave in a macho way.

I much prefer the Japanese concept, not least because Japanese companies have generally been much more successful than American— or indeed western—ones as a whole. The clear lesson, I believe very strongly, is that every possible opportunity has to be taken to maximise learning opportunities in an organisation. As in Japanese companies, we have to learn that time given to structured and shared learning is much more important than spending money on training.

Professor Kaoru Ishikawa, the late Japanese guru on quality, was really expressing the concept shown in Figure 4.9 in his quote at the beginning of this chapter. Customer satisfaction can only come from having the best possible quality in everything that your organisation does. And the only way to have the best quality is to have the most skilled (competent) people throughout the organisation. If you can provide better quality than your competitors, then you can charge a premium price—and that is the road to riches. Training therefore pays—and pays well!

This chapter contains many examples which may inspire you and your organisation to develop even better ways of learning.

6.1 The organisation as a learning environment

Potential learning from formal training courses is often much less valuable than it might be. The reason is simply that the environment at

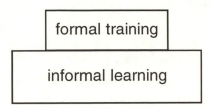

Figure 6.1   Is your training soundly based?

the workplace is not supportive of learning and the course itself is either not properly identified by the manager for its relevance to the employee or is not effectively followed up. As Figure 6.1 illustrates in very simple terms, formal and expensive training will only work effectively if it is based on a commitment throughout an organisation to maximising day to day learning. Therefore now ask yourself, both in terms of your organisation and for yourself: 'Is my training soundly based?'

In addition to the individual and management commitment to learning, as many ways as possible must be sought for learning and spreading expertise within the organisation. Every opportunity must be taken to pass knowledge and skills around within the company, whether verbally or written, one to one or in groups. Time has to be allocated to this activity, for this is the only way in which most organisations will successfully utilise the inherent talent and knowhow within, which gives them their commercial advantage over their competitors.

The failure of most organisations to maximise the use of their own experience and knowhow for growing and developing is because they fail to give priority to two key elements. These are:

1 *Time*—for example, one hour each week for each employee.
2 *Structure*—learning activities must be planned and controlled.

Some examples of useful practical activities which I strongly recommend are:

- lunchtime/evening/worktime seminars;
- team group meetings;
- presentations to staff after attending training courses;
- a brief summary document circulated with key information/lessons/action after a training activity;
- summary sheets of activities and learning from the previous week presented and circulated to colleagues;

- planned distribution of documentation;
- short-term secondments and projects;
- work shadowing;
- mentoring.

An example of the very effective use of just one of these, summary sheets of activities and learning from the previous week presented and circulated to colleagues, comes from an engineer who was seconded from a British company's research laboratories to Toshiba in Japan for six months. He described how every Monday morning each member of his team of 12 engineers had to summarise on a single sheet of paper everything which he had done and learnt during the previous week. He had just five minutes to explain this to a group meeting of all his colleagues and these sheets were then pinned up on the wall for everyone to refer to during the coming week. Thus, by the end of the first hour of work each week, every member of the team was fully up to speed on everything done and learnt by the whole team. Those who argue that Japan is so culturally different that there is nothing which the West can learn from it are closing their minds to some very simple lessons and techniques, such as this one, which any organisation can easily implement at very little cost.

In summary, for CPD to be as effective as possible:

- time must be committed in a structured way;
- money must be committed in an organised, planned and controlled manner;
- there must be organisational and individual visions of the future.

Let us therefore look at a series of case studies to see how a number of organisations, in different ways, have maximised their organisational learning.

6.1.1 Food for thought at Courtaulds Engineering

Courtaulds Engineering in Coventry has obtained considerable benefit from organising regular lunchtime seminars for its technical staff—typically for 20 to 30 people at each session. Lasting from one and a half to two hours, these have proved to be an economical and successful way of training and updating engineers and technicians in specific engineering topics.

Many of the seminars have been presented by technical representatives from outside the company—including suppliers, contractors and consultants. The topics have included noise, safety, equipment technology (such as boilers) and pressure systems regulations for the mechanical engineers. Electrical engineers have had seminars on subjects including variable speed drive systems, circuit breakers and aspects of the IEE Regulations, while the instrumentation engineers have had sessions in areas as diverse as ultrasonic level measurement and batch software. The process engineers also have their own regular programme of lunchtime seminars and the civil engineers and architects will soon be doing the same. Some lectures are also provided by Courtaulds' own experts. The company is also improving its design expertise through the use of the Institution of Chemical Engineers' video training courses.

Tim Harrison, principal engineer, says: 'We are using state of the art technology, so what better way is there of keeping our staff updated than inviting our suppliers to provide these seminars? There are also benefits to the suppliers in improving their understanding of what our needs are as customers.' Roger Burley, senior manager, engineering, adds: 'Engineering is not only about knowing the technology; it is also very important to know where to go for an answer and what to ask. These regular sessions help to extend our engineers' networks.'

'In addition,' says Frank Passey, chief engineer, electrical, 'by participating in these lunchtime seminars presented by external experts, they also experience a variety of techniques, helping them to make better quality presentations themselves. This is particularly beneficial to our graduate trainees. We also often require our engineers to make a presentation to their colleagues on returning from any formal training course.'

Other approaches used by Courtaulds to develop its technical staff include a training centre, providing open learning facilities including PC software and languages.

6.1.2 Building on success at Bechtel

Bechtel, with over 21 000 professional staff operating in more than 135 countries, is consistently ranked among the world's leading professional and construction organisations.

In the UK office there is strong support for CPD and involvement with professional institutions. For example, the company pays institution membership fees and there is a level of seniority above which technical employees cannot rise unless they are chartered engineers.

The company's commitment to CPD includes:

- individual career development plans;
- a technical track mentor programme for technical specialists;

- skills matrices;
- regular and wide-ranging in-company lecture and seminar programmes;
- overseas placements for many staff;
- providing experience and training in the UK over an extended period for engineers from Russia and Oman prior to opening up offices in those countries;
- recording and monitoring the precise numbers of fellows and members of professional institutions.

Riley Bechtel, president of Bechtel Group, Inc., says: 'Bechtel has a firm belief in the development of its employees. We are committed to training and development and follow a policy of promotion from within.'

As part of this policy, staff have the opportunity to study for the Bechtel Business Certificate, which teaches general management and business skills. This is run in conjunction with Middlesex University and successful completion can lead to exemption from the first year of the university's three year part-time MBA course.

Bechtel has been in the UK since 1952 and provides a full range of engineering, procurement, construction and project management services to the power, civil, petroleum, chemical and mining industries throughout Europe, Africa, the Middle East and Southwest Asia.

The company has been involved in some of the world's most challenging engineering and construction projects: supporting fire-fighting operations and the reconstruction of Kuwait's oil producing facilities following the Gulf War; assisting Eurotunnel in the management of the Channel Tunnel project, one of the world's largest construction undertakings; development of Qatar's North Field, the world's largest nonassociated gas field.

In recognition of its significant contribution to the export of UK goods and services, Bechtel was awarded The Queen's Award for Export Achievement 1994. This was Bechtel's second award in four years, having won it for the first time in 1991.

6.1.3 Investing in people—the prime resource

Oscar Faber TPA (OFTPA) was the first consultancy in the construction industry to be awarded the title Investor in People (IiP), which includes requirements for effective CPD.

IiP is a national standard implemented locally by Training and Enterprise Councils (TECs) in England and Wales, and Local

Enterprise Companies (LECs) in Scotland. An organisation wishing to gain IiP accreditation contacts its local TEC or LEC, which provides advice on how the company's CPD measures up against the standard. In some cases there may be some additional support in the form of consultancy to help with the preparation of an action plan. The company must also make a formal commitment to work towards the standard. When the organisation is ready, the TEC's assessors carry out a detailed investigation of what the organisation actually does to develop its employees.

In the case of OFTPA, the company already had all the necessary elements in place; these included:

- identification of individual and group learning needs;
- an annual programme for training;
- an appraisal system;
- planned learning, both on and off the job;
- assistance in obtaining qualifications;
- monitoring and evaluating training in the context of business objectives;
- communicating with all staff (and not just to them);
- promoting a learning culture;
- procedures for staff development as part of the quality assurance system.

However, it was felt necessary to send out a questionnaire to all staff to assess their perception of what actually happened. Nigel Lloyd, staff development manager, said: 'The feedback proved to be a real eye opener for us and enabled us not only to address the weaknesses which were identified, but also to strengthen our other evaluation processes.'

'Our commitment to staff development has shown many tangible benefits, most importantly in staff retention and in increased professionalism and flexibility,' says Richard Brown, assistant managing director. 'We have also worked with the Management Charter Initiative and Department of Employment on introducing the new national standards of management into our performance appraisals.'

The increasing number of organisations achieving IiP status ranges from, for example, Foster Wheeler, with 2400 employees, to the Institution of Chemical Engineers with 84 staff. Each is awarded a plaque, is allowed to call itself an Investor in People and becomes entitled to use the distinctive IiP laurel garland logo on all stationery and publicity. Normally accreditation lasts for up to three years.

6.1.4 A powerful approach to CPD

A powerful commitment by the senior management to developing staff at all levels has led to major benefits at National Power's Ironbridge power station.

The challenges provided by privatisation in 1992 initiated a review of the skills across the business. Senior management decided to adopt a major programme of training aimed at developing multiskilled employees, with the objective of empowering people throughout the organisation. Not only has the skills base been raised significantly, but responsibility is now delegated to the lowest practicable level. At the same time, the number of levels on the site has been reduced to four: the station manager, first line team leaders, foremen/supervisors and operators.

The benefits have included an overall increase in profitability of sixty one per cent, with productivity per employee up by 111%. Better efficiency in converting fuel to energy has resulted in savings of £550 000 a year and the plant has 95.18 % availability, compared to 89.97 % before the training. Furthermore, Ironbridge has set a new National Power safety record by clocking up 1000 days without a lost-time accident on the site.

The philosophy behind all this has been to raise quality levels by expanding knowledge across the site. Where specialist expertise is identified in particular employees, they are given off-line training in trainer skills. Their challenge then is to cascade that knowhow across the site to all those who will benefit. For example, 17 new information technology systems and programmes were cascaded in this way. In addition, supervisors have been encouraged to train for the National Examining Board for Supervisory Management certificate, which has now been achieved by over 250 staff across National Power; this involves residential courses, distance learning and work-related projects over 12–15 months.

Bill Yorke, ash and dust foreman, for example, is being fully empowered to enable him to run a self-sufficient team which can operate in its own right. This involves him in taking over a number of new areas of responsibility. 'It is beneficial to both the station and ourselves as life becomes far easier when all responsibilities are under one roof,' says Bill, who joined Ironbridge as a mechanical fitter 25 years ago. 'Because I am involved so closely in the area, I can see just where the money can best be spent, and perhaps other areas where it doesn't need spending. I will also be able to look very closely at specific jobs and see if there are other ways of doing them just as effectively but for less money.'

180 employees are now registered on National Vocational Qualification programmes of assessment. Operators are being given craft skills, with the craftsmen and fitters in turn being cross trained into new skills, including those of the operators. Those in functions such as engineering, logistics, finance, environmental services and human resources are developing a service culture, based on a detailed skills analysis, backed up by structured training plans.

Not surprisingly, Ironbridge power station won one of the 59 National Training Awards for employers in 1993. In addition, it won commendation from the Management Charter Initiative. The station has also been working towards Investors in People accreditation.

6.1.5 Moving towards a learning organisation in Hong Kong

In Hong Kong the 1990s have been characterised as turbulent years with companies facing a rapidly changing social, economic and political environment. This level of change is likely to escalate as Hong Kong heads into an increasingly uncertain future and a company's ability to deal with this dynamism is likely to be the chief arbiter of its success. In order to maintain and improve its high level of service to the Hong Kong community, the Mass Transit Railway Corporation (MTRC) believes that the key to long-term success is the creation of a learning culture.

The MTRC adopted a competency-based approach to staff management in 1990 to facilitate a culture of continuous improvement. The company developed a structured way of describing behaviour in the organisation, looking at the added value of managers and providing a common language for integrating both strategic and operational level personnel and business plans. The competencies critical for success at three different levels in management were agreed and, early in 1994, extended to supervisory levels.

A combination of objective and empirical methods was used to identify the key competencies critical for the success of the organisation. These methods included using structured computerised questionnaires, the repertory grid method and the critical incident technique (where managers are asked to identify the significant competencies which they have used in a number of different situations). Recognising the dynamic nature of Hong Kong, these key competencies are regularly reviewed.

Use of these key management competencies starts at the recruitment stage, ensuring that individuals are a best fit for their jobs. They are also integrated into the appraisal and performance management system, and used in designing training and development programmes. This has allowed the MTRC to reward those whose competencies best meet its business needs and to maximise its investment in its employees. It also provides a framework with which to identify the quality of the staff resources currently available, as well as to plan staffing needs over the next five years.

Assessment centres are used to assess groups of managers for possible promotion using well qualified observers in an off the job setting. The assessment tools used include tests, questionnaires and behavioural simulations.

The MTRC believes that organisational performance can only be improved by the commitment of individual managers to their own self development. Therefore all managers are encouraged, but not forced, to participate in the two-day assessment events. By the end of 1993, 19 executive managers (63 %), 59 senior managers (92 %) and 76 junior managers (33 %) had already attended these centres.

The assessment feedback report is discussed by the participant's manager, the observer, the participant and a development adviser, whose role is to advise on the appropriate next steps. The adviser continues to review progress with the participant until a development plan has been implemented, offering individually tailored advice and support at every stage.

However, responsibility for implementing the development plan rests firmly with the individual, supported by the line manager who carries out a final review approximately a year after the date of the assessment.

The MTRC believes firmly that CPD is the key to its future success. In 1994 the concept of competency was cascaded down to supervisory levels and a detailed analysis of the technical competency needs was also made.

A database has been set up to monitor all the data from the evaluations made at the recruitment stage, the annual appraisal and the assessment of potential at the development centres. All this is used for manpower and succession planning to support the organisation's future business plans.

6.1.6 Genesis at Harland & Wolff

In 1989, after privatisation, Harland & Wolff (H&W) in Belfast signed a two part agreement with Kawasaki Heavy Industries (KHI)—a design package for tankers and a technical assistance programme. As part of this agreement there has been open exchange of information, including that about personnel, training and development, as well as productivity rates.

KHI, for example, obtains a significant productivity improvement from its employees every year, even before any money is invested, by the approach to initial training and CPD. H&W's objective is to establish a continuous improvement environment and expect this to take four or five years. The objective is to be world class, with a clear vision and set of clear values. Project Genesis was launched to remodel all job roles in the company.

H&W staff have shadowed managers and directors in KHI at all levels and modelled their organisational structure on that of their partner company. So far, 160 H&W staff have worked with KHI in Japan for one to five weeks and some KHI staff have worked at H&W. George Spence, general manager of productivity development, is in charge of the Genesis programme. He says: 'We believe that a most important aspect is the role of middle managers, who provide the key link between the front line operations and senior management. We have therefore introduced development centres for general foremen and managers, based on assessing the key competencies required for their roles.'

The emphasis one year might be on team working and team building, while the next year the emphasis will move on to the development of the skills of coaching. Team leaders are responsible for developing the skills matrix in their area, with this information prominently displayed as part of a visual factory concept. Visitors can immediately see what the problems are, what is being done about them and how successfully.

In some areas, such as computer aided design, H&W is ahead of KHI. Over the period 1989 to 1993, productivity at H&W improved by fifty per cent, with another forty five per cent improvement by the end of 1994. All processes in the company are benchmarked against the KHI equivalent time and every effort is made to equal or better them.

Recently Masaaki Imai, the Japanese quality expert, visited H&W and commented on how impressed he was by the housekeeping standards, which he judged to be higher than in some European aircraft companies.

6.1.7 Going to town with the institutions

Lewisham Borough Council is facing new challenges as a result of compulsory competitive tendering; by 1996 ninety per cent of all work done in engineering in local authorities had to be put out to open tender. As a result of past commitment to training, the council sees itself as being in a particularly strong position to meet these challenges.

With the support of John Coach, the borough engineer, a new position of technical training officer was created in 1989, filled by John Maher. He established a policy of updating the expertise of the 48 engineering staff by:

- introducing a comprehensive training programme;
- encouraging all staff to join either the Institution of Civil Engineers (ICE) or the Institute of Highway Incorporated Engineers (IHIE).

A new training agreement with the engineering staff was drawn up and launched in May 1993, with presentations by Eric Jenkins from the ICE and Judith Walker from the IHIE. This followed a comprehensive training needs analysis based on interviews with all members of staff and their managers. A prime consideration was the improvement of customer service.

All engineering employees now became entitled to one day of training per week up to December 1993. A comprehensive programme of courses was scheduled throughout the year, which any engineering employee could apply to attend. These ranged from a two-hour session on 'Dealing with customers' to a two-day course on 'ICE conditions of contract'. To maximise the effectiveness, the programme included:

- a complete set of course notes for each session;
- up to half of each in-house course devoted to dealing with actual problems encountered in the workplace;
- job shadowing;
- provision of a library of books, publications, video and audio tapes.

6.1.8 Claro Precision Engineering

The Engineering Council has published a number of case studies showing how organisations have benefited from CPD and Investors in People (IiP). One of these case studies describes Claro Precision Engineering Ltd, with 34 employees.

Located in Knaresborough, North Yorkshire, the company was formed in 1978 with three working directors and a company secretary. The 14 000 sq ft factory is now equipped with CNC machining centres, turning centres and milling machines to provide a machining, assembly and toolmaking service on a subcontract basis. Claro has a particular expertise in producing tooling for medical implants.

The company is fully committed to a total quality management philosophy and its quality system is registered to BS5750 part 2/ISO9002.

Within a short time of setting up in business, Claro had established itself as a company with a good reputation for the delivery of a quality product which met the customer's specification. The demand for products rose steadily, but the directors were conscious of a poor financial return which showed few signs of improving. An initial investigation revealed that, although the product that left the factory would meet the customer's requirements, rework and error correction prior to acceptance for despatch was excessive. Existing management controls had failed to reveal these shortcomings and it was evident, too, that junior managers lacked the techniques necessary for management. The company's policy of internal promotion had not been accompanied by appropriate training and development. The Training Agency was contacted and a consultant brought in. (This developed into strong links with the North Yorkshire Training and Enterprise Council when it was formed in 1991.)

Every employee, including the directors, was given appropriate training covering a wide range of skills. It was found, for example, that the simplest of measuring and monitoring systems, if applied consistently by staff trained in their use, would bring substantial improvements—including a five-fold reduction in rework. The concept of simultaneous engineering was also introduced, with consultation between designers and production staff. Employees now had a wider range of skills resulting in a flexibility which allowed them to assist in all parts of the manufacturing process.

Turnover increased by twenty three per cent in 1993, with improvements in profitability, in spite of an unfavourable economic climate. There has since been a substantial increase in orders in hand and a twenty per cent improvement in on-time deliveries. Production costs have dropped through the improved control of the budget and a reduction in overtime working. The costs of cutting tools fell by fifteen per cent, and the cost of consumables fell by almost fifty per cent.

Claro was among the first companies to be awarded IiP status in the engineering sector in the UK. Howard Chadwick, Claro's managing director, says: 'Without this investment in training Claro would have suffered a declining share of the market which, along with our rising costs of production, might well have brought about the demise of the company, despite the acknowledged success of our products in engineering terms. Instead, Claro is very much in business and having to face up to a new problem—demand beginning to outstrip production capacity. Even here the problem is being tackled by staff who have identified improvements which will increase capacity without prejudicing the cost of production. A few years ago such practical and innovative thinking would not have been possible.'

6.1.9 BAA—continuous improvement takes off

BAA, which operates seven airports in the UK and has extensive international interests, has the mission to 'make BAA the most successful airport company in the world'. This means:

- always focusing on customers' needs and safety (over 500 000 passengers are interviewed annually and their views steer continuous improvements of all aspects of the business);
- seeking continuous improvement in the cost and quality of its services;
- enabling all employees to give of their best.

Change is fundamental to BAA's business, with a daily investment of well over £1 million. But, as Graham Matthews, project services director, says: 'Despite our success as a company, we still have many opportunities to explore in our quest to reach the standards which we believe are possible. For example, UK contractors are often only thirty five per cent productive, and we know that construction costs can be reduced significantly. Our continuous improvement (CI) programme is aimed at just these sorts of issues.'

This continuous improvement of the project process was initiated in May 1994 and led to a launch of new initiatives throughout the company in 1996. In parallel with this, a research programme in conjunction with Reading University and Warwick Manufacturing Group, is developing a world class project process, which will be adopted by the company following rigorous trials and testing. These two parts of the CI programme will be accompanied by a substantial training programme aimed not only at BAA's internal project management and engineering staff, but also at consultants, contractors and suppliers. The three key issues for BAA in this quest for world class standards are:

- construction processes in the UK are often as backward as processes in the UK motor industry used to be;
- the UK construction industry does not provide the same level of value for its customers as other industries or construction in some other countries;
- because BAA has such large purchasing power, there is an ideal opportunity to create lasting change.

Training for change

BAA has already run a series of internal training seminars to create heightened awareness of the change programme and to identify the key issues involved.

A detailed skills and needs analysis has been undertaken throughout the company, which has identified three key areas for those involved in construction projects:

- technical;
- business skills;
- interpersonal skills.

'A detailed training programme for BAA project management staff is being introduced,' says Graham Matthews, 'and at the same time we are developing an induction and training programme for our preferred suppliers. It is only by working with our suppliers to develop their products and services that we will be able to make the improvements that we need to make to achieve world class standards. We will be setting measurable targets and objectives for performance improvements.

'Our internal training will be largely based on interactive workshops. Management of design, for example, will be about looking at real life design issues and finding examples of good and bad design.'

Ensuring effective follow up

BAA is very conscious of the danger that much of this investment in training for change will not be followed up effectively.

Therefore, there will be considerable emphasis on training line managers to be effective coaches and mentors of their staff. In addition, staff will be encouraged to attend refresher courses to maintain the momentum of the programme.

6.1.10 Using CPD to move into the first rank

Rank Xerox Technical Centre (RXTC) in Welwyn Garden City has some 1000 high quality staff supporting a worldwide high technology business with a £3.5 billion annual turnover. The company's total commitment to improving even further both its customer and employee satisfaction is led by Dr. Guy Rabbat, executive director, who says: 'We have a very clear five-year vision—to build a virtual organisation. By the turn of the century our 1000 employees will be supported by another 5000 virtual employees. Other organisations sell equipment; we sell our people's skills and knowledge.'

A key part of Rank Xerox's vision is to form alliances with value added reseller partners (VARPs), enabling partnerships between Rank Xerox's technologies and skills and those of the VARPs.

The four corporate priorities in the company are:

- constantly monitoring and measuring the satisfaction of customers;
- using a wide variety of measures to evaluate employee satisfaction;
- maximising the return on its assets;
- increasing market share.

For the last three years the company has placed considerable emphasis on identifying the technical competencies required of each employee. Strategic thought has also been given to forecasting the migration of skills mix which will be required over the next five years, as is shown in Figure 6.2. To support employees in enhancing their existing skills and developing new ones, a wide range of sophisticated and innovative learning tools and technologies has been developed; these include competency, career development and job vacancies tools, computer based training and open learning.

'We are living in a culture of rapid technological change—moving from Rank Xerox the reprographics company to Rank Xerox the document company,' says John Green, manager of training and development. 'We must have multiskilled, multitasked teams, which have to be both adaptive and innovative.' The company has developed a competence dictionary of 435 competencies, each defined at five levels of proficiency, and is mapping these competencies to business performance.

To this end RXTC has developed a competency-based analysis tool—

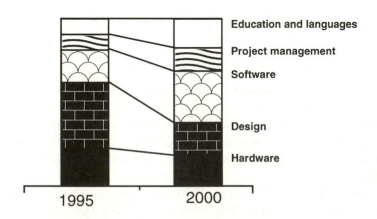

Figure 6.2   The changing skills mix in the Rank Xerox Technical Centre

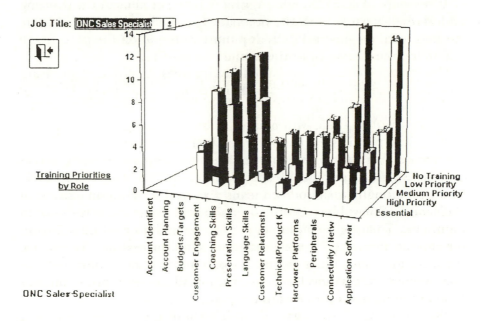

- Identifies areas of greatest need
- Promotes focused investment of training budgets

Figure 6.3   A screenshot from the Xerox Assessment of Competencies Tool

the Xerox Assessment of Competencies Tool, an example screen of which is shown in Figure 6.3. This allows continual monitoring of progress against the criteria critical to the successful performance of a role. Individual results can also be collated with other results to identify and prioritise both individual and group development needs.

'We believe strongly in empowerment and constant CPD for everyone,' says Dr. Guy Rabbat. 'Our managers are appraised in particular on the basis of how well they develop their staff. Any manager failing to do this well would not remain with us for long.' All employees attend a highly successful employeeship one-day event, designed to help them to release even more of their potential and ideas to support the company. 'We are now moving from empowerment to self-directed workgroups and then on to self-managed workgroups,' adds Dr. Rabbat.

Employee satisfaction is closely monitored. This has risen from forty nine per cent expressing satisfaction with the company in 1992 to seventy seven per cent in 1995; this compares with a national UK norm of fifty two per cent. In addition, ninety four per cent of the training provided is shown to meet the prescribed competencies. All managers are appraised on their staff development skills and on the percentage completion of identified staff training.

This rising success in developing employees is mirrored in rising customer satisfaction, market share and company efficiency. The level of customer satisfaction rose from sixty per cent in 1992 to eighty-seven per cent in 1995, with market share increasing from 54.9 % to 83.1 %. Return on assets has also risen by fifty per cent.

All this is overseen by an employee steering committee which, as one of its activities, encourages all employees to find themselves a mentor. In addition many imaginative means are used to communicate with employees. Everywhere in the organisation there are large colour monitors displaying a continually changing menu of information on everything from job opportunities to sales successes. These are supported by many other means of obtaining instant communication with employees, ranging from internal e-mail to Lotus Notes and the Internet, as well as a range of in-company magazines.

There are thus many ways in which there can be benefits from investment in CPD. We will look, finally, at one individual's viewpoint.

6.1.11 Gillian Sturke, CEng—engineering specialist, Northern Telecom Europe, Basildon

'Career development was very far from my thoughts when I responded to an appeal for female engineers to work as role models in local schools. Having recently become an MIEE and registered as a CEng I felt that my professional development was progressing well.

'As part of a liaison scheme between industry and schools being promoted by the Essex Education Authority, I spent as many afternoons as my business commitments allowed at a local primary school working with groups from a class of eight and nine year olds. The aim of my visits, as part of one of the topics, was to introduce the children to electricity and simple circuits. There were also broader aims—to increase their awareness of life outside the narrow confines of the classroom and to bring a new perspective to technology lessons, a subject which few primary teachers are trained to teach.

'As well as finding the project rewarding and enjoyable, I was surprised to find how much of a challenge it proved. It made me see a whole new side to the communication of ideas, as well as making me review my own specialisation.

'I hope that my experiences will improve my ability to communicate ideas and concepts, especially to audiences with less technical knowledge than my own, and in the motivation of my team and direction of my projects.'

6.2 Summary

There are, then, clearly a multitude of ways of sharing learning across an organisation and of developing your competencies and career as an individual.

Now take another look at your PDP and see if there are any ideas in this chapter which you want to build into either your own or your organisation's development plan.

Chapter 7

How to use mentoring and coaching as part of your CPD

*'Everyone has a bit of a natural coach within them'—Frank Dick, former
director of coaching, British Athletics Federation
'We must view people not as bottles to be filled, but as candles to be lit'—
Robert H. Schaffer*

I keep returning to Figure 2.1 and emphasising the importance of all three factors being in place for effective continuing professional development to take place. In the middle of this figure lies the concept of the manager as a coach; this role, I believe, is the glue which holds CPD together in any organisation.

In this chapter we will look at what this word coaching really means and how it relates to counselling and mentoring. How do you become a good and effective coach or mentor? Do you need to be qualified as a psychologist? What are the key competencies required? What is the link with appraisals? How can you make best use of coaches and mentors as a vital part of your CPD?

Let us start by looking at the word coaching and see how it links to the word management, particularly since it is my strong view that most organisations in the West have a quite erroneous idea of what management is all about.

The importance of this chapter is that coaching/mentoring is a universal skill. Even if you are not a mentor yourself, you should really find yourself a mentor as a key part of your CPD, so it will help considerably to be able to recognise the attributes of an effective mentor. Furthermore everyone, whether a coach/mentor or not in any formal sense, can use the approaches described here to persuade others of the merits of their ideas—whether it be their own staff, their colleagues, their boss or even their spouse!

Coaching/mentoring is the most important competence required not only of a manager in particular, but in addition of everyone involved in CPD.

7.1 What is coaching?

All too often, coaching is seen as an activity performed by someone who is an expert in a skill and who shows another person how to perform better at that skill. I do indeed know people who use their considerable business experience, for example, to coach individuals so that their performance is improved. However, I also know many people who very effectively and successfully coach people in areas of expertise in which they, the coaches, have little or no experience.

To explain this apparent anomaly, let us see the balances in the roles of a coach as shown in Figure 7.1. You will see that the coaching spectrum includes, at one end, putting things into people, like skills, knowledge and experience—which is probably the way you have seen coaching up to now. At the other end, however, it is also extremely effective at pulling things out of people, such as their potential, their commitment and their expertise.

In most organisations in which I have worked as either an employee or a consultant, the only real expertise that was needed to make it a world class winner was within the organisation itself. The amount of specialist knowledge needed from outside has usually been relatively

Figure 7.1   The coaching spectrum

little. The problem in most cases, however, in the majority of non-Japanese organisations at least, has been that expertise has not been effectively shared internally. Similarly, most individuals already have many years of experience in their area of specialist interest; despite this, all too often they have difficulty in pulling this expertise together in a way which maximises their own potential.

In both cases, then, of organisations and individuals, the need is for some external support or stimulus to help to focus and direct the expertise already available within. Words which readily describe this external supportive activity include coaching, facilitating and mentoring. As Figure 7.1 indicates, the spectrum for the coach, facilitator or mentor ranges from the coach doing most of the talking (retaining high control of the discussion), to the coach doing most of the listening (giving the control of the discussion, to a large extent, to the person being coached).

Too much of management activity, in my experience, is on the left of this spectrum. How many managers, after all, do you know who are good listeners? Probably all too few. And do you respond better to managers who tell you what to do, or to those who listen to your ideas?

Assuming that you agree that listening is extremely important in a manager's set of key competencies, then we have a good starting point for deciding on the skills required of a good coach.

7.2 What is mentoring?

Mentoring differs from coaching in two key respects:

1 The mentor needs to be more experienced in the areas of expertise concerned than the individual being mentored.
2 The mentor should not, generally speaking, be in a line management position above the person whom they are mentoring. This is both because that person needs to be able to confide in the mentor to a degree which is usually difficult to achieve with one's own manager, and also because it provides a secondary source of advice, which can be seen as independent.

The usual context in which I am involved with mentors is in my activities with the professional institutions. Most, if not all, institutions insist that for a training scheme for young technicians or engineers to be accredited, each trainee must have a mentor. The definition that I use of mentoring is: 'The use of experienced employees to accelerate the learning and development of those less experienced.'

Some of the key roles of mentors in their relationships with the individuals being mentored are to:

- facilitate professional development;
- provide positive influence;
- influence colleagues to provide and maximise learning opportunities;
- share experience;
- provide feedback.

However, it is increasingly being realised that mentoring is not just a role that is important at the beginning of someone's career. Rather it is highly desirable that we have one or more mentors throughout our careers, and mentoring is therefore an extremely important aspect of CPD. I myself have had two or three mentors at any one time, including one at present who is the vice-president of the Japan Society for Engineering Education.

It is worth emphasising that mentoring is not just highly beneficial to those being mentored; it is also very valuable CPD for the mentor. Benefits to the mentor include, for example:

- satisfaction in playing a key role in someone else's development;
- self development through learning how to be a more effective mentor;
- influencing successors;
- developing new perspectives through being challenged by, and challenging, those being mentored.

The sorts of factors that you should take into account in choosing mentors include their:

- background experience;
- interpersonal skills;
- motivation to succeed in the role;
- interest in helping less experienced colleagues.

Give some thought now, therefore, to whether, if you are not already one, you might not become a mentor as part of your own CPD. In addition, how might you find yourself a mentor? Mentoring is being recognised increasingly as a key part of CPD, both for the mentors and the people whom they are mentoring.

What is management?

Disillusion with the way in which organisations are run has been a problem for a long time. In the last century Mark Twain, the American writer and humourist, said:

'If work were so great, the rich would have hogged it long ago.'

More recently, the American quality expert Phil Crosby wrote in his book 'Quality is free':

'I worked for the company for 12 years before I realised managers were there to help me. Before that I thought they were a punishment from God.'

In Chapter 2, I described hearing the chairman of Sony, Akio Morita, speaking in 1981 at the IEE and criticising Peter Drucker for not understanding what management is all about. 'In Japan a manager's role is very simple; it is to develop the skills of his staff so that they can find better ways of satisfying the customers,' said Morita. That was the first time that it had occurred to me that management was really all about coaching and developing staff. So novel does this appear in the West that I have yet to come across any MBA course (and I have obtained the brochures of every MBA course run in the UK recently) which makes any mention of the skill of coaching in its brochure.

And yet I have still to meet anyone who disagrees with Akio Morita's view of management as predominantly being about the skill of coaching. Nor, in many presentations in many countries around the world, have I met anyone who disagrees with the concept that organisations should look like the one in Figure 2.5 and not like the one in Figure 2.4.

Indeed, one of the most successful business books in the 1990s, 'Reengineering the corporation', points out that:

'While a manager can typically supervise only about seven people, he or she can coach close to thirty.'[10]

Unfortunately the authors failed to realise the prime importance of this statement and are thus apparently still puzzled as to why seventy per cent of business process reengineering initiatives fail; it is because the processes of Figure 2.1 are not in place—and above all because managers are not behaving as coaches.

I have been doing a consultancy project in Turkey with a colleague, Gordon Williams, managing director of Gestalt Consultants. In essence the consultancy assignment is aimed at enhancing the coaching style of management, which has enabled this company to be the best in its class in the world. Part of our work has been coaching the management on a one to one basis, and encouraging these directors, managers and supervisors to coach their staff in turn also on a one to one basis. The objective of this is to release as much potential and as many ideas from all employees as possible. In one case the maintenance supervisor had

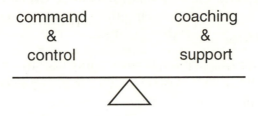

Figure 7.2   *Balancing the management style*

been challenged by the production director to reduce the time taken for changing two large reaction vessels from 18 days to 12 days. By coaching each of his team on a one to one basis and obtaining their ideas and commitment, the maintenance supervisor succeeded in completing the task in just six days, which was subsequently further reduced to just three and a half. This is one of many such experiences that I have had of the enormous power of coaching.

It must be emphasised, however, that it is not always appropriate for managers to act as coaches the whole time. As Figure 7.2 illustrates, there is a range of management behaviour between command and control on the one hand, and coaching and facilitating on the other, and a balance is needed between the two. There are unquestionably times when managers must behave in command and control mode— for example when safety is important and clear rules must be specified and action taken if anyone transgresses them.

If you are a manager or supervisor, take a little time now to ask yourself how much time you spend at each end of this range of management behaviours. List the types of situation in which you have been commanding and controlling, and those in which you have been coaching and facilitating. What does this analysis tell you about your management behaviour?

In my experience, most managers generally fail to behave as coaches for much, if any, of their time for one very simple reason. They have never been effectively coached themselves and therefore do not understand its extraordinary power. They therefore give coaching neither the importance nor the priority that it deserves. They do not feel themselves to have the competence to coach their staff, and they do not feel committed to coaching. This chapter is therefore aimed at making you feel very much at home with the concepts and giving you the confidence to develop your skills in this area.

As Frank Dick was quoted as saying at the beginning of the chapter, everyone is a natural coach to some degree. One of the barriers in my own mind used to be the idea that I needed a formal qualification in psychology to be able to coach others. I now know from direct personal experience that that is quite untrue. Indeed, in all my experience I have never yet met anyone who is not, to some degree or another, a natural coach. On the other hand, we can all become better at coaching—and the best way to do this is to constantly practise the skill and to obtain as much feedback as possible.

If coaching is indeed the prime, although not the only, skill required of a manager, then mentoring can be seen as a subset of that prime skill. A mentor uses much the same skills as a coach, but with the important difference noted earlier; specialist knowledge about the area in which you are coaching is not required, whereas in mentoring the opposite is generally true. For example, the IEE insists that anyone mentoring someone on an IEE accredited training scheme should be a chartered engineer.

7.3 Coaching/mentoring is about lighting candles

As the other quote at the beginning of this chapter suggested, we can see the people we are coaching or mentoring as being 'like candles to be lit, rather than as bottles to be filled'.

One context in which this might be considered is shown in Figure 7.3. Spend a few moments and think through where in the four quadrants your recent discussions with others at work have been. Have they been looking at problems or opportunities, at the past or the future?

You will no doubt have marked a range of crosses mentally in the four quadrants. If you have many in the top right hand box, then you

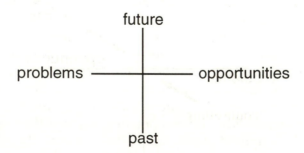

Figure 7.3   Where are conversations held?
       Mark the area of the model in which your recent conversations have
       taken place

are either illustrating that you are a natural coach or else you are fortunate enough to have colleagues or a manager who have the right approach.

Lighting candles is about encouraging others to think positively about the future. Coaching and mentoring are very much about helping people to focus their plans on future opportunities. In contrast, the other word used frequently in this area is counselling, which is also an important skill. Generally speaking, the role of a counsellor is to help people to overcome problems and tends to be about events in the past. As Figure 7.4 shows, there is therefore an important difference between counselling on the one hand, and coaching and mentoring on the other. Clearly, there is some overlap between them; for example, sometimes in order to help someone to focus on the future, you need to help them to come to terms with, or learn from, the past. But a key skill in coaching and mentoring is this ability to focus on the future. In all three cases—counselling, coaching and mentoring—the range of skills required is very similar, but the contexts tend to be different.

It is also important to realise the very wide range of situations in which coaching can be valuable. You can coach your staff at work, for example. But you can also coach your boss; indeed, the most dramatic payoff that I have encountered, when working as a consultant helping organisations to develop their skills in coaching, came within half an hour of the end of one of my workshops when one of the delegates coached his boss so successfully that there was a Canadian $1 million payoff within three weeks—a story which I will tell later in this chapter.

Figure 7.4   The difference between counselling and coaching

Equally, you can coach your colleagues. Or you can use the skills of coaching to sell your ideas or products inside or outside your organisation.

You can coach groups. And you can coach yourself, which I frequently do when I have a problem or an opportunity.

7.4 The key competencies of a coach/mentor

You might like to take a sheet of paper and write down the key competencies which you believe to be the most important for a coach/mentor.

Those I and my colleagues have found to be particularly important are shown in Figure 7.5. This list is, of course, not exhaustive, and the skills required will depend to some degree on the particular context. You may wish to change the list for yourself as your experience of coaching/mentoring others develops.

The most important skill of a coach/mentor is the ability to listen actively. Listen very carefully to what the other person is saying. But do much more than that. Hear how they are saying it. What does their intonation tell you—the pitch, pace, tone and volume of the words

active listening
questioning skills
giving praise and recognition
rapport building
creating trust
being nonjudgemental
being candid and challenging
ability to work from the other person's agenda
giving encouragement and support
focusing on future opportunities
getting to the point
observation skills
being objective rather than subjective

Figure 7.5   The key competencies of a coach

that they speak? Watch their face carefully and note the micro-movements of their facial muscles. What is happening to their eyes? Observe their body posture. What can you deduce from their body movements? Are they relaxed, with their arms in a comfortable position, or are they nervous or defensive with arms folded, etc.?

Next, concentrate on asking the right questions. Engineers tend to ask questions starting with *what...?* This is an excellent type of question, obviously, but do note that it is very task oriented. In addition, you need to ask process questions, often starting with *how...?* The danger here is that engineers, by concentrating too much on tasks and not giving enough attention to the processes required to achieve them, often fail to meet their objectives. On the other hand, by concentrating on getting the processes right, you will often find that the tasks almost look after themselves. Therefore, when coaching or mentoring, do put plenty of emphasis on exploring with the other person how they either went about solving a problem or plan to do so in the future. What could they have done better with hindsight? Where might they take the project next or find the information they need? When will they implement their plan?

Two bits of advice which I find are generally useful are:

- try to avoid using the *why...?* type of questions. These tend to make the other person rather defensive and to invite self-justificatory answers. Rephrase them by asking, for example, 'How did that happen?' etc.;
- one measure of success that you can use as a coach or mentor is to observe how much of the time you are talking in relation to how much time the other person speaks. All too often, when running workshops in this area, the delegates all agree that listening is important, but when they then actually start practising their coaching skills on a one to one basis with a colleague, they find that they are not only speaking themselves for most of the time, but, even worse, telling the other person exactly how to solve their problems. Bad habits die hard, so concentrate on this important skill of active listening. By asking the right questions, and steering the other person towards finding the right or best solution for themselves, the effect will usually be very much more powerful than if you revert to the command and control style of management.

You might also find it valuable at the end of meetings with your colleagues or your staff to ask the question: 'How did you feel about that meeting?' This is a very good process question, and the feedback will help you to become ever better at your process skills.

Work through the list of other key competencies of a coach from Figure 7.5 and ask yourself how good you are at them.

Do you give sufficient praise and recognition? I remember some years ago accompanying one of our company employees who had just won the Young Woman Engineer of the Year award back from the high profile national presentation and asking her how things were going back at work. 'You know,' she told me, 'in all the years that I have worked for this company, no-one has ever thanked me for anything that I have ever done.' Sadly, I keep hearing too many similar stories, so do please remember to thank your staff, colleagues, boss, clients, etc., when appropriate. As they say, praise does wonders for the sense of hearing.

Do you establish the best possible rapport with people, for example by noting their body language and emulating it subtly? Do you create a situation over time in which people can readily trust you and can you force yourself to be nonjudgemental, so that other people can explore their own worlds of opportunity, success and failure under your expert coaching?

Can you judge the right time to be candid and challenging to the other person?

Do you take as your starting point, when coaching, what the other person wants to be the starting point, rather than forcing your own agenda down their throat? What do they really want to talk about? What are their real concerns at the moment? You can usually only find the answers to these questions by asking them!

Do you give the right encouragement and support when they need it, and can you help them to focus appropriately on the future opportunities open to them?

Do you get to the point reasonably quickly when coaching or mentoring, and do you use your observation skills of the type of language they use, the pallor of their complexion and the nature of their body language?

Can you encourage them to be objective in their analysis of situations and can you be objective in helping them?

Finally, two other bits of advice. First, do not be afraid of silence. If the other person lapses into silence, you may well have asked an extremely effective question, and they may need a little time to think through their response. If they stop to think and look upwards/sideways/downwards this is often a sign that you have asked something which is causing them to think very carefully. Secondly, if you consider that it really will help the other person to offer your own view or solution, preface it by asking their permission to do so, for example, by asking: 'Do you mind if I offer you an idea?' This helps the person to move their mindset into a situation where they are open to your ideas.

Now rate yourself in terms of your current level of competence against each of the key competencies listed below:

	Current level of competence					
	poor				excellent	
active listening	1	2	3	4	5	6
questioning skills	1	2	3	4	5	6
giving praise/recognition	1	2	3	4	5	6
rapport building	1	2	3	4	5	6
creating trust	1	2	3	4	5	6
being nonjudgemental	1	2	3	4	5	6
being candid and challenging	1	2	3	4	5	6
ability to work from the other person's agenda	1	2	3	4	5	6
giving encouragement and support	1	2	3	4	5	6
focusing on future opportunities	1	2	3	4	5	6
getting to the point	1	2	3	4	5	6
observation skills	1	2	3	4	5	6
being objective rather than subjective	1	2	3	4	5	6

Now ask yourself which of these key competencies you wish to concentrate on improving. Furthermore, ask yourself how you will know that you have improved. What will your measures of success be?

It is important to note that these coaching skills need not be restricted to formal coaching or mentoring sessions. You can make use of them informally in your day to day interactions with others, including over the telephone and when passing them in corridors, etc.

Now look back at the work that you have done in designing and filling in your PDP. Amend and develop your PDP in the light of the self analysis which you have just done on your coaching key competencies and write in the targets for self improvement on which you wish to concentrate.

7.5 The cycle of success

We will now look at the first of three very simple models of human behaviour which you will almost certainly find useful in coaching or mentoring others.

The cycle of success illustrated in Figure 7.6 shows the complete cycle which we need to go through in order to complete a task or achieve an objective successfully. It is another example of the way in which it is important to look at things from a process point of view.

Going through the cycle may take a matter of minutes or may take several years. In every case, however, you will need to have an idea, create a plan to implement it, and decide to implement it. Some ideas never get beyond this point, for whatever reason!

On the other hand, the value of many engineers is in their ability to generate new ideas. When I was in Hong Kong for the IEE on a training accreditation visit, I was told a very interesting story by the director of a very well respected international engineering company. He explained how a recently deceased senior member of the company had been kept on the payroll prior to his death despite having suffered a stroke and being significantly incapacitated. The company continued to pay him a high salary on the basis that he had the much valued ability to come up with just one brilliant new idea each year.

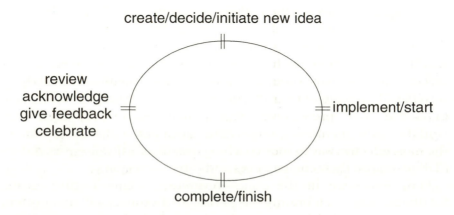

Figure 7.6   The cycle of success

In the cycle of success, having made the definite decision to go ahead with a project, there will then come a time when you press the button and the implementation actually starts. It may be surprising sometimes how easy it is to become a little stuck at this point. One of my own favourite devices for avoiding making a critical telephone call, for example, is to go and make a cup of coffee. Or you may be tempted to double check everything before you start, or to seek approval or advice from yet more people. The role of a coach is to help the other person to understand where they are on the cycle of success, to see what the barriers are to moving a project around it and to find ways of moving around the cycle faster. See it as being rather like a flywheel; the faster it spins, the easier it is to keep up the momentum of your activities.

The next stage in the cycle is to complete or finish the project. Here engineers are notorious for seeking ever greater perfection in a design before they agree that it is complete. Too many companies have been bankrupted by such engineers, since no design is ever perfect other than in terms of being fit for the purpose intended; the urgent need may be to get the product in a good quality state, but maybe not fully developed, to the marketplace. Mentors have a very important role here in helping young engineers, fresh out of university, to understand the complexities and trade offs necessary in finalising a design or project.

Another key reason why people tend to fall down at this point in the cycle is because they are afraid of being judged; once they announce that the project is complete, then their manager or their peers may quite possibly judge them to have failed. Much better, some people think, to postpone the evil day when this might happen. Your role as a coach/mentor is to check whether this, or any other reason for failing to complete a project at the earliest opportunity, is a hindrance.

Yet another possible problem may be that the person concerned lacks the necessary skills at the right level. Again, by asking the right questions and by being aware of the full range of that person's competencies and career objectives, you should be able to coach them as effectively as possible. How much of their PDP they are prepared to share with you is, of course, their prerogative; even if they wish to keep their PDP confidential to themselves, they will be much better able to respond in the most effective way to your coaching questions if they have used their PDP to analyse their competences and career objectives.

The next stage in the cycle—reviewing, acknowledging, giving feedback and celebrating—is critical, because all too often, particularly in task-driven engineering organisations, the pressure is to move straight to the start of another task once one is complete. This is bad practice for a number of reasons:

- it is exhausting to move straight from completion back to start;
- it is very important to review how the project went, decide on what might have gone even better and consider how to apply the lessons learned to future projects (both for yourself and for others), etc.;
- you need to acknowledge success and give yourself, and others, credit for the achievements;
- it is highly appropriate to celebrate successes and give yourself the extra energy to start and complete the next task even better and even faster.

Therefore, as a coach or mentor, do ensure that this stage in the cycle is not overlooked.

Now take a little time to identify two or three projects on which you are currently working. Where are you at present on this cycle of success? What is the next step? What are the potential barriers and how might they be overcome?

In the case of mentors, it is very important, I believe, to base the mentoring sessions around the PDP or logbook used by the person being mentored. Ensure that at the front of the document there is a copy of the overall training programme for the two years, for example if it is an IEE accredited training scheme. This is the first step in the cycle, ensuring that you both have a clear overview of the training plan. Does it cover all the areas or elements required by the institution, by the company and by the trainee? For each training experience, both on the job and off the job, ensure that there are clear written and agreed objectives before-hand, identifying the elements of the training scheme which this particular training experience should satisfy. At the end of each training session, ensure that the trainee reviews and writes up the experience in his or her PDP; as the Mentor, you should then discuss this, clearly identifying which elements of the training programme have been successfully completed and where there are still gaps to be plugged—and how. Write your comments in the PDP or logbook as a record for the future—of what has been successfully learnt, or possibly misunderstood; treat the PDP as a learning document with an audit trail of the learning process. Put your initials in the PDP as part of this audit trail and date them. An example of a well completed IEE PDR sheet is shown in Figure 7.7. (One benefit of the IEE system is that at the end of their two-year training experience, graduates can simply photocopy the relevant section of their PDR—the ones called 'Structured formation training'—and send it off as the key part of their end of training XIT form, thus

Dates		No. of weeks	Description Aims/Objectives	Trainee's report/Mentor's comments	Training Elements	Mentor's Initials
From	To					
Nov 9X		(1 day)	**ELECTROMAGNETIC COMPATIBILITY COURSE** A 1-day course run by the Industry Research Association to give awareness of the importance of EMC in Design.	This short technical course highlighted for me the disastrous consequences of not taking EMC into consideration at the design stage of a project, and gave useful information on how to design to minimise EMC problems.	C2	*A.N.G.* *20/12/9X*
Jan 9Y	Apl 9Y	13	**MANUFACTURING & PLANT ENGINEERING** To work as a full member of the electrical section of the Plant Engineering Department. To become fully conversant with the PLCs (Programmable Logic Controllers) used in the factory so as to be able to specify PLCs for new equipment, and solve PLC-based problems. To look after the Department's responsibilities for H&S. *& to gain a good working knowledge of H&S! A.N.G.*	Organised trials of a number of PLCs from different manufacturers and drew up a detailed check list against which they could be compared. Acted as Plant Engineer responsible for the Electrical Repair Centre. Represented the Department at Health & Safety meetings, and took follow-up actions. This placement gave me the first-hand experience of real engineering. *Engineering is often as much to do with people as it is about things! I will recommend to Training that you are nominated for the course on "Effective Teamwork" which you should find very useful now.* *A.N. Ginear CEng. MIEE 18/5/94*	C1, C4 C6, HS	
Apl 9Y	July 9Y	12	**APPLICATIONS IN ENGINEERING & TECHNOLOGY** (AET) at A Local College of Further Education The course was aimed at giving engineers practical experience in designing and making a piece of electromechanical equipment. The areas covered by lectures and practical workshop periods were: Production Methods Workshop Skills Materials Engineering Drawings & Product Documentation Design & Specification Production Management Project Control & Finance CAD, Manufacturing & Engineering	As I have an electrical background, this was an ideal opportunity to develop my metal-working skills. I made sure I became familiar with casting, turning, milling, grinding, drilling and welding. I designed, built and tested the power and speed control circuits for the spindle speed controller, and also the rev. counter. The closed-loop control circuit allowed the speed of the permanent magnet DC motor to be varied for different materials, and then kept constant for varying loads. Before this course I did not know how to design and produce PCBs (printed circuit boards). It has allowed me to gain valuable experience and skills in practical techniques that complement the theory learnt at University. Now I have used these skills to help produce a complete computer-controlled machine tool.	B	*A.N.G.* *7/7/9Y*

Figure 7.7   An example of a well completed PDR sheet

saving a lot of time and effort filling in the details of their training and obtaining their mentors' signatures.)

And do not forget the celebrations!

Use of the cycle of success by both mentors and those being mentored using a PDP to focus the learning is a good example of applying the overall concept of Figure 2.1 effectively. It is also the basis on which all institution initial training schemes are based, and they work very well, forming an excellent start to young people's careers. However, the danger is that things do not then continue in the way in which they started. CPD is a lifelong learning concept and the principles which work well at the start of a career should apply for the rest of one's career as well. Therefore, do ensure that part of your CPD consists of finding yourself one or more mentors, as well as yourself becoming a mentor to one or more other people.

7.6 The change model

The change model is one which is universally powerful to anyone in any situation where they are trying to persuade others, quite apart from its use in coaching and mentoring.

When you are in fact coaching or mentoring someone, whether in a formal situation or informally, you will frequently find that they are having to face up to some sort of change. Alternatively, you may be helping someone else who needs to persuade a colleague, a client, their boss, or whoever, to change. As we all know, most people are rather conservative at heart and persuading them to change can be extremely difficult.

Here, therefore, is another very simple model, the change model, which has been widely used and valued. It is shown in Figure 7.8 and has two axes; on one axis are the past and the future, while on the other are the internal and external dimensions of a situation.

Ask yourself what is likely to go through your mind if someone suddenly suggests a change to you, such as having to move your office. It is highly probable that you will seek an anchor in the past and see it as some sort of external threat. It is also quite likely that your initial reaction will be one of denial—denial that the change is either required or possible. This change model was originally developed for counsellors who were helping clients to overcome traumas, such as the loss of a parent. The initial reaction for most of us faced with the news of a parent's death is to deny that it is possible. That just seems to be the way

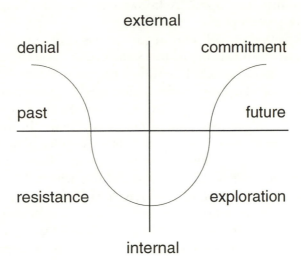

Figure 7.8   The change model

our minds work, and the value of the change model is that it helps ourselves and others to understand better how to deal with change.

Faced, then, with change, we need to recognise the importance of giving people the time to absorb the news or implications. In the case of bad or difficult news they may need the time to grieve. For less traumatic news of something new happening, it still requires time to internalise the new idea or concept. One example I remember from long ago was my A-level physics master—an excellent teacher—who always introduced a brief outline during the last five minutes of a physics lesson of the concepts which he was going to teach us in the next day's lesson. His theory was that we then 'vottled' (a word he invented, but which never made it into the Oxford English Dictionary) the new concepts overnight and were better able to understand the next day's lesson quickly. With hindsight I now realise that he was in effect using the change model.

In addition to giving people the time to get to grips with the change, you may also need to be quite firm in making it clear to them that the change is inevitable and will happen. You might suggest ways in which they might handle the change or explain what is expected of them.

Given time, most people will then internalise the need or inevitability for the change to occur. At this point, it is likely that they will create some form of resistance, and your role as a coach is to ask questions which steer them through this resistance. Ask them, for example, how they feel about the change. Try to bring their views and feelings out into the open. Acknowledge their feelings about the change and show empathy.

If all goes well, they will then begin to cease hanging on to the past and status quo, and will begin to explore the possibilities. Ask them questions such as *what if...?*. Encourage them to picture in their minds the possibilities and opportunities which might open up for them as a result of the change. Be as supportive as possible and encourage them to target specific objectives. Help them to decide on the key issues and the next steps which need to be taken. Have regular planning sessions with them to plan and review the progress on these specific steps. Explore the sort of new experience or training which they might need.

Eventually, all being well, you will have coached them into becoming committed to the change. Give them as much encouragement and credit as possible, and celebrate successful use of the change model! Also start discussing with them the longer-term objectives which now need to be targeted.

Again, like the cycle of success, this is very much an example of how placing emphasis on the process by which change is achieved is far more likely to lead to success than pure concentration on the task itself would be likely to do. The classic failure to implement change successfully is as a result of forcing change too fast, as shown in Figure 7.9. Far too many managers and engineers try to move from the denial quadrant straight into the commitment quadrant, with the result that people react strongly and negatively and move into the resistance quadrant; shifting them out of there is then very much

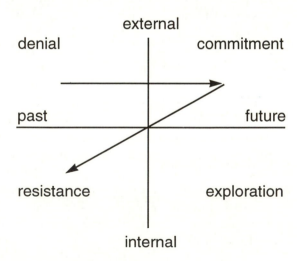

Figure 7.9   The effect of forcing change

more difficult than if the change had been handled more effectively in the first place.

An interesting example of the successful use of this change model, referred to earlier in this chapter, was with a client where I was providing consultancy support as part of a culture change process. A colleague and I were running a workshop on coaching and, as part of the discussion around the change model, one of the delegates explained that he had been having a problem persuading his boss to send him to Canada to follow through some problems with a sub-contractor. He said that every time he raised the question of going to Canada with his boss, the discussion turned rapidly into a confrontation. This delegate, who was a very senior manager, decided to use this challenge as the practice coaching exercise following the explanation of the change model. As it happened, he had a meeting with his boss just half an hour after the end of the workshop. He used the meeting to raise the Canadian issue again, but this time was less confrontational, and mentally used the change model. He gave his boss time to internalise the need to take the issue up directly with the contractor (denial). By asking appropriate questions he discovered that much of his boss's opposition to him going to Canada was the time it would take (resistance). He therefore suggested that he flew out there over a weekend to minimise the time he was away, and in return his boss started exploring the possibility that while out there, another problem could be resolved with another contractor (exploration). The net result was that, twenty minutes after the meeting started, this senior manager came out of his boss's office, literally skipping in the air (celebration!), saying: 'I'm going to Canada!'. Three weeks later he returned from Canada with a $1 million cheque from the contractor—an excellent example of the potential benefits of upwards coaching!

Note that this manager was in no formal sense mentoring his boss. He was, however, informally coaching him—with very powerful results!

7.7 Comfort/stretch/panic

As a coach or mentor one of your roles is probably to raise the game of the other person. Another useful little model to have in mind around this purpose is Figure 7.10.

The two axes here are a person's perceived ability to achieve a task, and the perceived difficulty they see themselves facing. Low difficulty

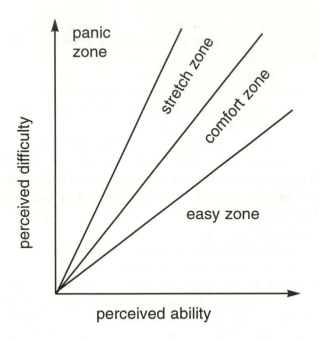

Figure 7.10   Balance between perceived difficulty and ability

and high ability clearly makes the task easy. Greater difficulty, but reasonable ability makes it a comfortable objective. Greater difficulty still and rather less perceived ability creates a situation in which someone becomes stretched. If the perceived difficulty rises even further, with the perceived ability becoming less, then a stage will be reached at which panic starts to take over.

The value of keeping this model in your mind when coaching or mentoring is that you can check where the other person is in the spectrum. Ideally they should spend as much time as possible in the stretch zone. At one end, they might have too many easy or comfortable tasks, where they learn very little and their PDP is likely to remain blank. At the other end, they will lose some of their self confidence, not to mention sleep, if they enter the panic zone. One excellent way, of course, of avoiding entering the panic zone is to develop the necessary competencies well before the challenges for which they will be needed have to be faced. Since it may take a long time—perhaps even several years—to develop these competencies to the necessary level, it shows the importance of applying the CPD principles outlined in this book. One important value of a mentor is to encourage the person whom they are mentoring to take a longer term view of the need for self development.

7.8 Linking to appraisals

As Figures 4.7 and 4.8 show, most appraisal systems fail to work because they are not based on continuous competence and career development.

Mentors and coaches, particularly managers using their competencies as coaches, have a vital role to play in supporting employees to develop their competencies in a way which best helps them, on a continuous basis, to achieve their job performance targets—and ideally even exceed them.

One interesting development in recent years has been the idea of upward appraisals, where employees complete appraisal forms each year giving feedback on their manager's performance. Usually the upward appraisal forms from all of a manager's staff are collated by a third party—typically from the organisation's personnel department—and then fed back to the manager, thus enabling each employee to give honest views. Often this process is reinforced by a meeting between the manager and all his staff, facilitated by the third party, to discuss the feedback. How much more effective, surely, all this would be if the employees were encouraged to develop their individual and group upward appraisal skills on a continuous basis, rather than just once a year.

7.9 Summary

This chapter has been aimed at providing you with ideas and techniques which you can use to improve your coaching and mentoring skills. All the evidence is that everyone has natural ability in this area, but the best way of getting even better is by having the confidence and commitment to practise and improve these skills on a regular basis.

The payoff will be very simple, your ability—as Professor Charles Handy describes it—to raise both your own and other people's potential from perhaps twenty per cent to something very much higher. The rewards are great, the risks are small, so go for it!

Chapter 8
How to make the best use of software for CPD

'One of our former US presidential cabinet members projected that by the year 2020 information will double every 73 days!'—James B. Appleberry, president, American Association of State Colleges and Universities

The use of software has been referred to a number of times already in previous chapters. There are so many different ways in which computers and the associated software are impinging more and more on our professional and personal lives that anyone who is computer illiterate will be increasingly handicapped from a career point of view.

Therefore, ensuring that you have at least one personal computer, and ideally two (an office PC and a laptop) is likely to form a very important part of your CPD. The amount of hard disk memory that software requires these days means that it is important to invest in the largest possible machine which you can afford. I myself have two 66 MHz, 500 Megabyte hard disk office PCs, as well as two laptop PCs. One of my office PCs has stacker software which increases its hard disk capacity to 1.2 Gigabytes, and yet it is already eighty per cent full. I have had to invest in a second laptop PC simply because, after two years, the 25 MHz speed of my first one proved to be too slow.

Some uses to which you might put your PCs to develop your CPD are:

- developing your keyboard skills and using wordprocessing packages, including for writing articles and even books;
- keeping details of all your business, professional and personal contacts using a database. It is said that President Clinton has kept a book of the names and addresses of everyone he has met throughout his career, and that he has used this to advance his political career. How much easier life would have been for him if he had been able to keep all this data on a software database!
- using spreadsheets to keep track of your finances, for example;

- accessing information on CD-ROMs;
- using personal development plan software;
- using groupware such as Lotus Notes;
- connecting into the Internet.

Let us now look in more detail at how these applications might all form part of your CPD.

8.1 Developing your keyboard skills

There are many software packages available to help you to develop reasonable keyboard skills. Given the amount of use of the keyboard that you are likely to need to make, it is foolish to stay as a two-finger typist. For the sake of a few hours of dedicated keyboard skills self training, you will soon develop the right keyboard technique using all ten of your fingers and thumbs. The best way, as with any skill, to improve further is by continual practice. Indeed, if you do not continually practise this skill, then your competence and confidence will decrease.

With Windows and Apple software, modern wordprocessing packages are relatively simple and intuitive to use. Do make the effort to become reasonably expert at one package. Target the time and commitment to do this in your PDP and give yourself credit as your skill level progresses.

Doing your own wordprocessing has enormous advantages. In my case, for example, in writing this and other books, I can sit down at my PC terminal whenever it suits me and can easily update and modify the chapters. I can import diagrams and charts from other software packages such as Excel, Powerpoint or CorelDraw.

Having my own hardware and software also enables me to produce my own diagrams for my overhead projector slides. When I write course notes for the many workshops which I run around the world, or write articles, newsletters or books, it is very simple to import these diagrams into the text. I now have approaching 1000 diagrams in my software and am continually adding to them whenever I have a new idea or hear or read something interesting. I can access all this material very easily on my laptop PC when travelling or in meetings. I can readily access my database to look up names, addresses, telephone numbers, fax numbers, details of past meetings or conversations on the telephone.

So many businesspeople now travel with their laptop PCs in their briefcases, that at least one major international hotel chain, Marriott, is designing rooms specifically to deal with this need. According to Marriott's brand director Leslie Cappetta, between 25 and 30 per cent of the business travellers interviewed as part of a survey at US airports now carry a notebook computer. All too often some physical gymnastics is required to identify a place where you can plug in your power and modem leads in your hotel room. In response to this, in 1995 Marriott signed a partnership with the American telecommunications giant AT&T and the furniture maker Steelcase to design the hotel room of the future. Prototypes were due to appear in the USA, London and Hong Kong over the rest of 1995. Sockets will be built into custom-made desks which can be moved around the room. Lighting will be improved to help work with a computer, and high speed data links for videoconferencing will soon become standard.

This also emphasises the importance of ensuring that your seating position and the background lighting are suitable whenever you are using your PCs. There have been many cases reported of people developing repetitive strain injury, with crippling effects on their hands, through having adopted inappropriate postures at a keyboard over a period of time. Other reports suggest that people who make exclusive use of notebook PCs, with their rather cramped keyboards, develop pains in their hands, arms, back or shoulders. Do, therefore, invest in an office PC for your main software work, and ensure that you are adopting an appropriate typing position. Most keyboard training packages will give you advice on this; otherwise ensure that you obtain sound advice from an experienced source—perhaps your secretary!

8.2 Using databases for your CPD

Keeping up to date records on a database of your existing and possible new contacts can be an enormously powerful way of enhancing your CPD. As they say, it is not so much what you know, as whom you know, that matters in your career. Maximise that resource which far too many people greatly undervalue—your network of contacts.

Very often you face new problems. As Bismarck once remarked:

'There is no point in learning for yourself if you can learn from others and avoid the cost of experimentation'

(I accessed this quote, by the way, by looking up a slide which I made in CorelDraw after hearing a speaker quote it at a conference in Vienna last year.)

One of the important uses of personal development plan software which we will discuss later in this chapter is as a corporate skills database. In a very similar way, you can use database software to develop your own sources of expertise to which you can refer when necessary. People are a very important source of such expertise, including the way in which they can say: 'I don't really know the answer to that question, but why don't you try contacting the following person...?'

Typical details of your personal network which you can enter in your database are:

● name;
● address;
● telephone number;
● fax number;
● areas of expertise;
● source of information;
● date last contacted;
● general notes.

Be aware, however, that once you start keeping details of people on a software database then you need to register this, if you are in the UK, with the Data Protection Registrar under the legal requirements of the Data Protection Act.

Other information for which you can use a database might include details of books and articles which you have read. Alternatively, you could record these details using personal development plan software.

8.3 Using spreadsheets for your CPD

One way in which I use a spreadsheet is to keep track of the progress of my savings and investments. I can then take a realistic view of how much I can afford to invest in developing my consultancy business, how much time I can afford to spend writing books and articles, speaking at conferences, etc., and how much time I need to spend on fee earning consultancy activities. I prefer to live life in the round and keeping track of how my finances can support me to do that is very important.

Another use that I made of my spreadsheet software was when in 1994 I was asked to be the prime organiser of a senior industrialists dinner in the IEE South East Midlands Centre area, which was hosted by the IEE's President and had the IEE's Secretary and one of the Deputy Secretaries present, as well as a number of other IEE secretariat and local centre committee members. The spreadsheet was invaluable in keeping track of who had been invited, the acceptances, apologies, names of their secretaries, telephone numbers, etc. The evening itself, and the groundwork which I needed to do beforehand, were certainly a very valuable CPD experience for me!

8.4 Using CD-ROMs for your CPD

The next chapter will explain how CD-ROMs could be used as part of an open learning programme—which might form an important element in your CPD.

If you can afford the £100 or £200 extra needed to include a CD-ROM drive as part of your PC, then this will give you experience of using a technology which is likely to be of increasing benefit to you. CD-ROMs can store so much data that more and more material is being packaged and sold in this format.

Increasingly conference delegates, for example, are being given a copy of the conference proceedings in the form of a CD-ROM instead of as a set of printed papers.

Providing technical abstracts on CD-ROM is also increasingly common. To give just one example, the Institute of Marine Engineers and British Maritime Technology Limited released their 'Maritime technology abstracts' on CD-ROM in 1995. This makes available the complete contents of each organisation's bibliographic database on a single CD-ROM, with the additional benefit of powerful search and retrieval software. These CDs are available on an annual subscription with updates every six months. The two databases that comprise the 'Marine technology abstracts' contain over 50 000 abstracts, dating back to 1940, and are drawn from all major engineering publications. Together they capture the field of marine technology in an extensive range of bibliographic references including:

- marine engine design;
- offshore technology;
- marine safety;
- naval architecture;
- pollution from ships;
- shipboard management;
- computers in the marine industry.

The CD-ROM's management software offers users two search modes—simple or complex. Either of these can be refined using Boolean operators—AND, OR, NOT—for a more specific response. All fields in the databases are indexed and abstracts satisfying a particular enquiry can be found in seconds from any starting point. The full text of any abstract can then be provided through the rapid document delivery service, managed by the Institute's Marine Information Centre. The CD-ROM also offers a notepad facility for making notes against records and a bookmark facility for marking records in order to return to them at a later time.

8.5 Groupware—the most exciting learning medium for the future

Groupware is a new form of software which has begun to appear since the late 1980s and is particularly associated with the field of knowledge management. So what is it?

With most software a problem arises when using parallel copies of text, databases, spreadsheets, etc. For example, I myself keep, like many people, copies of these types of software on both my office PCs and my portable notebook computers. The problem arises if I do not update the notebook files from the office PCs before leaving the office and then add some data to the files on the notebook computers. I then either have to overwrite one set of files with the other, losing key data perhaps, or have to laboriously compare the files and update one set of data or the other.

Now this problem need no longer exist. The most popular and powerful form of groupware is called Lotus Notes* and uses a process called replication. In replication two text documents, databases, spreadsheets, etc., each of which has been separately updated over a period of time, 'talk' to each other. But now, instead of one set overwriting the other, the two sets add to each other all the data which they have had entered since they last replicated with one another. For example, I have written with two other authors a book called 'Upside down management—revolutionising management and development to maximise business success'[2] using Lotus Notes. When one of us was in Hong Kong and one of the others in Denmark, we could still add

---

* Lotus Notes is a registered trade mark of Lotus Development Corporation.

our new chapters to the overall book and both see what the other had added since the last replication to our respective office PCs from our notebook computers in our hotel rooms.

The power of groupware is enormous. It is not only invaluable in the mobile scenarios described above, it is also extremely powerful as a shared knowledge and learning technology for organisations of all sizes, including those with fixed locations. Even a small group of people in a shared office would find it immensely beneficial. At the large end, one major organisation, for example, has 65 000 licences for its employees around the world and uses it for many purposes, including preparing documents in parallel between offices in different continents and exchanging information and ideas with clients.

An excellent book which describes Lotus Notes in detail is 'Using Lotus Notes 4'[11] published by Que Corporation.

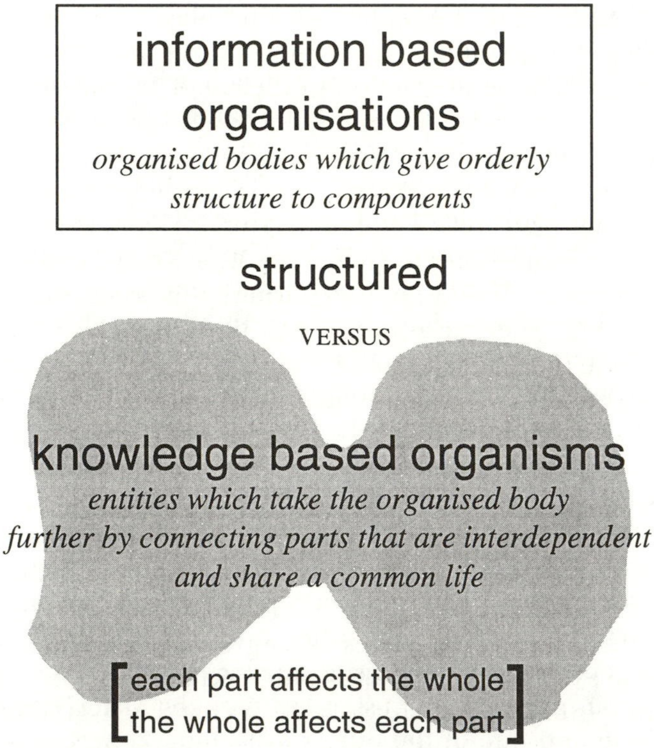

information based organisations
organised bodies which give orderly structure to components

structured

VERSUS

knowledge based organisms
entities which take the organised body further by connecting parts that are interdependent and share a common life

[each part affects the whole
the whole affects each part]

Figure 8.1   We prefer to work in organic environments since we are all organisms ourselves. Changes in each part affect the whole and vice versa

8.5.1 People are organisms

Organisations are generally rather rigid, linear, structured and information based. In contrast, organisms are flexible, holistic, fluid, dynamic and knowledge based.

One of the important insights that I have received from Ron Young, one of my co-authors on the upside down management book, is the one illustrated in Figure 8.1. As this shows, it is much more natural for us all to work for organisms, since we are organisms ourselves. In this way, changes in each part can affect the whole and vice versa. This is precisely what groupware allows you to do.

8.5.2 Using groupware for CPD

The implications of using Lotus Notes for CPD are enormously exciting. In many ways, by enabling large numbers of people to learn almost simultaneously from each other, Lotus Notes breaks the boundaries of space and time. For example, all the major six management, consultancy and accounting organisations in the world have copies of Lotus Notes for a large proportion of their staff. If someone in one of their Australian offices has a problem or spots a business opportunity, then he or she can enter that information into one of their Lotus Notes networks. If a response is not received from an Australian or New Zealand colleague before the end of the working day, then someone in the company's Hong Kong or Singapore office may be able to find a solution or exploit the business opportunity in some way. European colleagues, when they start work, may find that their Hong Kong colleagues have found a solution to a problem which they did not even realise existed, and they too are able to exploit this. Next, American colleagues wake up and on starting work they as well are able to maximise the opportunities initiated by their Australian associate. Therefore, by the time the Australian wakes up and returns for the next day's work, his or her problem or business opportunity may have been exploited around the rest of the world. If one sees all this learning through the job itself as a key part of one's CPD, then the opportunities for maximising CPD opportunities using groupware are mindboggling!

The simple example given above could perhaps have been dealt with through the Internet, which we will discuss later in this chapter. However, Lotus Notes is much more secure (it meets USA Department of Defence security criteria) and much more powerful in several respects. The process of replication is not inherent to the Internet, for example, (although one can link Lotus Notes software across the

Internet) and one use which organisations make of this Lotus Notes process is for preparing bid documents in parallel across one or more continents. IBM was so impressed by the potential of Lotus Notes that it purchased Lotus Development Corporation for $3.5 billion in June 1995. In January 1996 it released a new and even more powerful version, Lotus Notes 4, which was even easier to use, had much greater functionality, including an integral web browser, and—most importantly of all—reduced the price of a desktop licence from some $500 to just $50. Up until then just over two million Lotus Notes licences had been sold; however, IBM's marketing plans with Lotus Notes 4 are to increase that user base to 12 million by the end of 1996, to 20 million by the end of 1997 and to 30 million by the end of 1998.

Above all, using these very powerful forms of software can be enormously fun—and fun needs to be a key part of our CPD!

Even universities are now beginning to make serious uses of groupware. New York University, for example, has a virtual college, based on Lotus Notes, with each student having a PC and a modem, dialling up the central Notes server for lectures, with interaction with the tutor and other students being via e-mail.

One of the very significant and powerful aspects of Lotus Notes 4 is how closely it integrates with the Internet. Its format is very similar, it includes an integrated web browser and users can have a seamless link between Lotus Notes 4 and the Internet.

One of my major areas of work is developing means of using software to improve CPD. My colleague Ron Young and I have

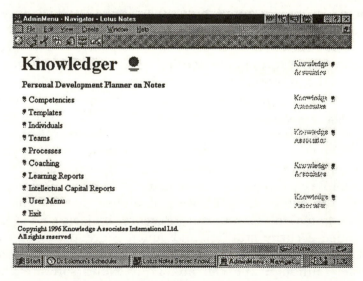

Figure 8.2   The Knowledge Associates PDP on Notes administration screen

developed personal development plan software based on Lotus Notes—called PDP on Notes—which we have launched through Knowledge Associates, the international knowledge management company. The administration screen is shown in Figure 8.2.

8.6 Using personal development plan software

In Chapter 5 we looked in detail at how you and your organisation might develop personal development plans (PDPs) and make the best use of them. At the end of the chapter, I suggested that some very exciting opportunities are now opening up for using software versions of PDPs, with the following advantages:

- linking learning experiences directly to your competence set;
- monitoring your learning to ensure that there is a balance between your short-term and long-term, and technical and nontechnical learning;
- using the powerful reporting facilities to monitor how much time you are spending on learning, whether you are obtaining learning linked to all of your key competencies, and so on;
- using the software as a skills database in an organisation.

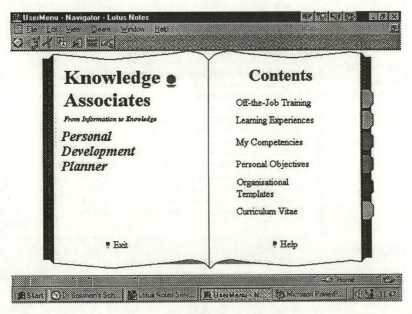

Figure 8.3   The Knowledge Associates PDP on Notes main user screen

A major objective of Ron Young and myself has been to make our PDP software as user friendly as possible. We therefore developed software which actually looks like a paper-based PDP, as shown in Figure 8.3.

Throughout the PDP on Notes software package, every effort has been made to enhance its ease of use and simplicity. By making the users feel that this is their own PDP, their commitment, enjoyment and satisfaction in using it is enhanced as far as possible. Key parts of the software include:

1   A section in which to record off the job training experiences (Figure 8.4). This enables the users to record in turn:

   • the date of the learning experience;
   • the time spent on the learning experience;
   • a description of the learning experience;
   • how they will benefit from and follow up the learning experience.

They can then tag each learning experience to one of their key competencies. Within that key competence they can then tag it to one of the subcompetencies, and then, within the subcompetence, they can tag the learning experience to a particular element. If they wish, they can also tag a particular learning experience to more

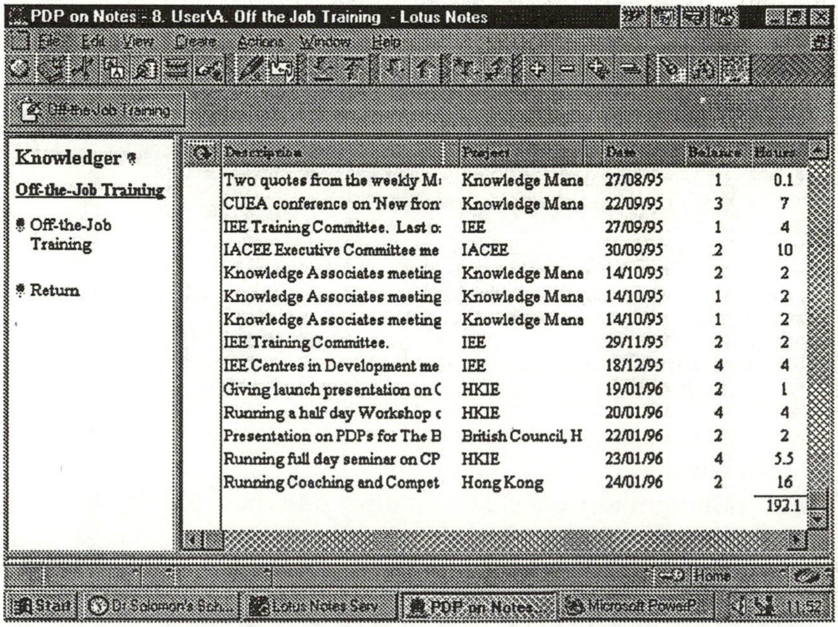

Knowledger	Description	Project	Date	Balance	Hours
Off-the-Job Training	Two quotes from the weekly M	Knowledge Mana	27/08/95	1	0.1
	CUEA conference on 'New from	Knowledge Mana	22/09/95	3	7
⬧ Off-the-Job Training	IEE Training Committee. Last o	IEE	27/09/95	1	4
	IACEE Executive Committee me	IACEE	30/09/95	2	10
	Knowledge Associates meeting	Knowledge Mana	14/10/95	2	2
⬧ Return	Knowledge Associates meeting	Knowledge Mana	14/10/95	1	2
	Knowledge Associates meeting	Knowledge Mana	14/10/95	1	2
	IEE Training Committee.	IEE	29/11/95	2	2
	IEE Centres in Development me	IEE	18/12/95	4	4
	Giving launch presentation on C	HKIE	19/01/96	2	1
	Running a half day Workshop c	HKIE	20/01/96	4	4
	Presentation on PDPs for The B	British Council, H	22/01/96	2	2
	Running full day seminar on CP	HKIE	23/01/96	4	5.5
	Running Coaching and Compet	Hong Kong	24/01/96	2	16
					192.1

Figure 8.4   The Knowledge Associates PDP on Notes off the job training screen

than one key competence, subcompetence and element. This section provides a vitally important learning audit trail, which validates each user's competencies on an ongoing basis.

Users can show whether this learning experience has mainly benefited their:

- technical/short-term development;
- personal/short-term development;
- technical/long-term development;
- personal/long-term development.

2 A section in which to record day to day learning experiences. This offers a screen almost identical to that shown in Figure 8.4.

3 A screen showing key competencies, subcompetencies and elements. It is these to which the users tag their off the job training and day to day learning experiences as described above. A typical such screen was shown in Chapter 4 in Figure 4.3, including:

- key competencies;
- current and target levels of competence;
- the name of the person who has authorised these competencies and levels;
- the dates on which each competence needs to be reviewed;
- a single flag if the competence is within 30 days of expiry and two flags if it has already expired.

4 A section in which users can record their personal objectives. These might well include appraisal objectives. It might also be where individuals target and record their successes in more personal objectives, such as improving a sporting performance or developing a hobby.

5 A set of screens which users can access to see which key competencies are required for other jobs within their organisations. In this way, they can see the sorts of gaps which there are between their existing competencies and those required in jobs that they might wish to do in the future. This has two important implications:

(i) users can target their career development that much more effectively;

(ii) it is much more likely that they will be able to focus their learning activities directly not only on their present jobs, but also on future jobs. Since the competencies they will eventually need for senior technical or managerial jobs may well take several years to develop, the sooner they target this aspect of their career development the better.

One very important advantage of using software for PDPs is that you and your organisation can monitor your skills and career development much more effectively. Many different types of feedback can be provided in the form of screen or printed report forms. As every engineer knows, you only get control in any system when there is feedback—and this principle is every bit as important in a learning process as it is anywhere else. Examples of the reports which can be generated regularly are:

- a detailed view of the organisation's key competencies, sub-competencies and elements for each employee;
- a list of the key competencies only of each employee;
- a list of each of the organisation's key competencies, listing under each one who in the organisation has that particular competence. In other words, this can be used as a skills database to search out and locate employees able to contribute a key competence to solving a particular problem which the organisation might face. This type of report can also be used for staff planning, to identify the current stock of skills in the organisation and compare these to those likely to be needed in future years. The corporate learning processes can then be directly focused on bridging the gap, thus maximising the organisational investment in learning and training;

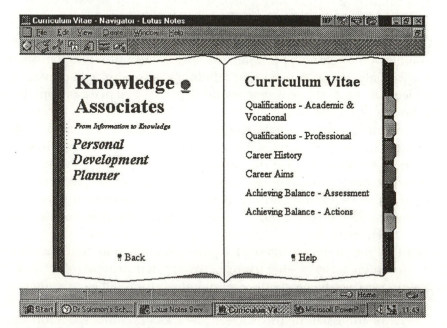

Figure 8.5   The Knowledge Associates PDP on Notes curriculum vitae screen

- reports on the key competencies, subcompetencies and elements which are either about to expire within the next 30 days, or else which have already expired;
- a list of all the key competencies, subcompetencies and elements identified across the organisation.

As Figure 8.5 shows, there is also a curriculum vitae section as an important part of PDP on Notes. This is also designed in a coaching style asking, for example, which have been the user's three most important learning experiences—and for what reason?

8.7 Have you really valued your organisation or personal assets?

For how many companies does their share price really reflect the human capital within them?

Instead, in almost every case, the share price and annual company report are filled with financial statistics on cash flow, turnover, profits and capital investment in new equipment. In the future, I confidently predict, these share prices and company annual reports will show the total stock of competencies within the organisation, with a value attached to them. After all, how else does an organisation provide the customer satisfaction other than in the way illustrated in Figure 4.9— by underpinning customer satisfaction with quality, which in turn must be based on competencies? The real bottom line in any organisation consists of the competencies and commitment of every employee.

Financial analysts will increasingly base their valuation of a company's share price on this total stock of skills, plus something which is even more important—the rate at which the skills base is growing. Therefore, it will become more and more essential for organisations to be seen to be using software such as PDP on Notes. Associated with this, we will see more and more sophisticated approaches to focusing individual and corporate learning on increasing this corporate skills base at the fastest possible rate. No longer will there be an obsession with the financial systems and controls alone; these will remain important disciplines and measures, but competence development, directly targeted at improving quality and customer satisfaction, will become of even higher importance.

Skandia Assurance and Financial Services has attracted much publicity in this area. Leif Edvinsson, for example, has the job title of director of intellectual capital and the company has:

- developed a complementary accounting taxonomy for intangible assets;
- incorporated critical success factors as quantifiable ratios;
- measured the speed of competence development and sharing as intellectual capital ratios;
- designed a model to provide a balanced picture of financial and intellectual capital.

Another organisation developing its thinking in this direction is The Dow Chemical Company, where Gordon Petrash has the job title of global director, intellectual asset management. He describes the goal of knowledge management within the company as being 'to add value to the business through the effective accumulation and use of intellectual capital'.

In PDP on Notes, there is an intellectual capital reports section (Figure 8.6), which allows the user to measure directly the value of the company in terms of such parameters as:

- the total stock of key competencies in the organisation, measuring both the number of competencies and their total score, taking into account the level at which each person's key competence has been accredited;
- details of who has which key competence at Level 3, the ability to perform the competence consistently;
- details of who has key competencies at world class level.

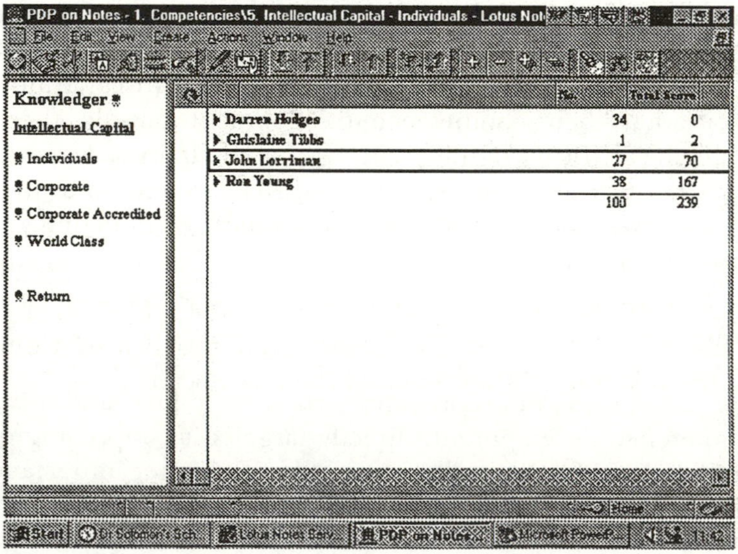

Figure 8.6   The Knowledge Associates PDP on Notes intellectual capital reports screen

PDP for Notes also allows organisations to monitor and measure the competencies of teams of people, which is a very powerful facility for focusing the CPD of individuals into achieving successes through teams.

Anyone wishing to try a free limited period version of PDP on Notes can download it from the Knowledge Associates' world wide web home page on www.knowledge.stjohns.co.uk. A Lotus ScreenCam demonstration version can also be downloaded.

One other extremely exciting development in Lotus Notes is a version called Domino, launched in August 1996. This allows access to Lotus Notes applications over the world wide web. PDP on Notes can therefore be made accessible to anyone with a web browser, without the need for those people to have Lotus Notes loaded onto their own PCs. Professional institutions, in particular, have shown great interest in PDP on Notes using Domino servers, since they will be able to:

- provide access to PDP on Notes to all their members who have web browsers, irrespective of where they are in the world;
- build up a database of the competencies of their members;
- observe the development of existing and new competencies amongst their membership and thus better target their literature, courses, conferences, evening meetings, etc.;
- build up template competence sets for typical engineering roles, based on an ongoing analysis of their members' competencies. This will allow members to compare their existing competencies with those typically required in other roles to which they might aspire. They will thus be able to focus their CPD much more effectively. In addition, they might well obtain considerably wider perspectives on how they can broaden their competence development outside their own discipline; electrical engineers in various disciplines, for example, could not only look at typical competencies required in other electrical disciplines, but also in aeronautical engineering, plastics technology, etc.

In January 1997 the IMechE and the IEE piloted a PDP on the Web. Active involvement in this also came from the British Computer Society, the Institute of Materials and the Law Society.

8.8 CPD using the Internet

Few people reading this book will not have heard of the Internet. Its origins go back to 1969 when the University of Utah, the Stanford

Research Institute and the University of California at Santa Barbara and at Los Angeles linked themselves using an experimental computer network funded by the Advanced Research Projects Agency (ARPA) of the US Department of Defense (DoD). This network was named ARPAnet and was designed to be sufficiently flexible to withstand a nuclear strike by having alternative routing options for the software being exchanged across the system. By 1972, when the network was made public, 50 universities and research facilities were connected, all of which were involved in DoD projects.

Towards the end of the 1970s other networks were appearing, some of which were funded, like ARPAnet, by the US Government, with others being private.

By the early 1980s these systems started coming together and were named the Internet, with the explosive growth really occurring in the early 1990s. No-one knows the total number of users (since no-one owns the Internet) but it is generally agreed that already tens of millions of people use it and that the number is growing by several million more each month.

The output is equally impressive: in the 18 months up to July 1995, users created three million pages of information, entertainment and advertising.

Until the 1990s, the Internet was accessible only to computer experts who had mastered the Unix operating system and were willing to write endless command lines in order to reach obscure destinations. The creation of icon-driven interfaces like Mosaic and Netscape in 1993 by Marc Andreessen and Eric Bina made it easy for anyone with a mouse to navigate through a welter of choices.

In order to access the Internet you need a PC with a modem, preferably of at least 9600 baud—otherwise exchanging data will take a long time and give you large telephone bills—connected to a gateway. In general, there are two types of gateway. First, there are gateways which only give you direct access to the Internet; one example is Pipex, which is used by IEE staff. Secondly, there are network providers, the largest of which is CompuServe, to which I am connected, and which also provide access to the Internet; you can either use these network providers to exchange electronic messages (e-mail), files and data with others connected to the network, or you can use the network to access the Internet.

Everyone on the Internet has an e-mail address, which identifies them uniquely to the Internet system. My own address is 100524.106@compuserve.com, which tells the Internet computers around the world to send my messages to the CompuServe computer based in the USA. Since there is a CompuServe telephone number at my local telephone exchange, which connects me to the USA automatically every time I dial it, I only have to pay local rate telephone charges when I send and receive Internet messages. The e-mail address of Ken Smith, the IEE's career development manager, is ksmith@iee.org.uk. In this case all the Internet computers know that any messages addressed to Ken need to be sent to the UK and then to the organisation called 'iee'. It may only take minutes for an e-mail to travel from one side of the world to the other, routed by a series of computers which pass it from one to another, choosing an alternative routing if necessary. Typically you pay a few pounds each month. (I actually have a second e-mail address of jlorriman@iee.org.uk, since the IEE has implemented a software system which arranges that any e-mail sent to this address is automatically rerouted to my CompuServe e-mail address.)

Much media hype has been written about cyberspace and surfing the Internet. This really refers to the world wide web, which otherwise is referred to simply as the web. This was originally developed starting in 1989 at the CERN particle physics laboratory in Geneva by Tim Berners-Lee and Robert Cailliau, with the purpose of helping scientists to exchange and share information. The first line mode web browsers appeared in 1991 allowing users to read pages as well as the cross references between pages, often on different web servers. A portable version of this software was released in the form of freeware in 1992 and by the end of that year there were just 50 Web servers. The first graphical Web browsers appeared in 1993, the most successful being Mosaic developed at the National Center for Supercomputing Applications (NCSA) at the University of Illinois. By the end of 1993 there were 250 web servers around the world, rising to over 2000 by the end of 1994. Now, of course, the growth rate in web servers is exponential, with the total almost certainly exceeding 100 000 and hundreds being added each day.

A major reason for the success of the world wide web has been what is called hyper text mark-up language, known as HTML; this allows a single description of a page to be displayed on a variety of client machines, and for pages on the web to be annotated with a number of tags which link to other pages on either the same or different web browsers and which control the appearance of the file document when viewed on screen using a web browser. The latest web browsers, such as Mosaic, Netscape and Microsoft's Navigator, are very user friendly software packages in the Windows software environment. One of the very powerful facilities in HTML is the way in which these tags can be used to define links to additional, related, material; this could be text or nontextual (such as sound, animation, still pictures, etc.), or other Internet sources on other computers elsewhere in the world. This ability to hop from one computer to another using a very easy graphical interface is what has caused so much interest around the world. Such is the rapidly increasing use of the web that one estimate in 1995 was that all Internet traffic was due to exceed telephony worldwide by the end of 1996.

Some 70 per cent of all new Internet subscribers are from companies, despite all the publicity given to individuals surfing the Internet. Businesses are increasingly using the Internet to exchange information with customers, suppliers and other contacts. Companies such as Fiat, General Motors and BMW provide information and demonstrations of their products on the world wide web. Other companies, including DEC, use the Internet to invite suppliers to submit tenders. Almost a third of British companies were connected, either directly or indirectly, by the end of 1994 according to an in-depth survey carried out for Pipex.

The American analyst Killen Associates estimates that transactions on the Internet will reach £380 billion by 1999. For example, the Royal Bank of Scotland has for some time been running a successful home banking system.

The number of opportunities that you have to enhance your CPD using the Internet is endless. You can, for example, browse through more than 750 000 book titles on the Internet Bookshop. This makes it possible to find texts when you have only the author's name or part of the title. Some 20 British publishers have provided contributions to this Internet version, including reviews, contents, jacket cover graphics, author biographies and even, in some cases, complete chapters. You place your order on screen and have the option of paying by credit card. All web addresses start with the format http://www (http means hyper text transfer protocol) and you can access the Internet Bookshop at http://www.bookshop.co.uk/.

The IEE itself has not been slow to become involved. It provides a wide range of information on the Internet, including details of:

- publications;
- books;
- distance learning materials;
- events;
- staff information;
- board and committee structures;
- INSPEC;

- membership;
- information on CPD.

In 1997 a CPD providers' database on behalf of many UK engineering institutions became available through the IEE on the Internet. All these services can be accessed using a web browser, such as Mosaic, Netscape or Navigator, at the address http://www.iee.org.uk/. There are well over 1000 pages already accessible, with many more becoming available all the time. The IEE web home page has been accessed by Internet users all around the world. In turn, the IEE's web pages link to many other sites.

Give thought, as well, to participating in Internet newsgroups, of which there are over 5000, and each of which is dedicated to a particular subject.

Be aware, however, of the significant dangers of computer viruses. I write this only hours after examining the first diskette that I have encountered with a virus; my wife was alerted to it when she tried to use this diskette as part of her job in a government agency. It had been nowhere near any of my machines when, yesterday at a government site, she put it into a PC running anti-virus software and was warned that it contained a boot-sector virus (later identified as the very common Form Virus). I do strongly recommend, therefore, that you use anti-virus software, particularly of the variety, like Dr Solomon's Anti-Virus Toolkit or Norton Anti-Virus, which runs continuously in the background and which detected my wife's infected diskette yesterday.

In summary, therefore, the Internet provides some very interesting and exciting ways to develop your CPD. One final example is at the University of Southampton, where students on the BSc degree in information engineering became the first in Britain to study for a degree using the Internet. They are able to attend lectures and tutorials without ever setting foot outside their homes, if they so wish. From their terminals they can download essay assignments and lectures. They can even attend electronic tutorials in which all the group log on simultaneously and hold an online discussion. Four days of face to face teaching is included, but all the rest of the course is over the Internet. The first participants included eleven mature students at the IBM research centre at Hursley in Hampshire. It is also possible now to study for an MBA in the same way, at Southampton Institute, which can be accessed at http://www.cecomm.co.uk/silcourses/mba-internetinfo.htm/; this is a part-time MBA with the help of online tutors.

If you wish to read more about the Internet, then you would do well to read the book 'The Internet for scientists and engineers' [12] by B. J. Thomas. Specifically written for the engineering and scientific community, this is a concise, thorough and clearly written, yet entertaining, guide.

8.9 Looking to the future

Are there any limits, one wonders, to the use of software to help develop one's CPD? New technology is simultaneously reducing the size of hardware and increasing the power of both hardware and software. One prediction that I make is that well before the year 2000 we will all have personal digital assistants (PDAs). These will be electronic terminals small enough to fit into our pockets, just as electronic personal organisers do now. However, these PDAs will have inbuilt modems, cellular communication telephone links and software for networking, such as Lotus Notes and Internet access software. In addition, they will have PDP on Notes software loaded. When new ideas or information come your way, you will be able to enter these not only into your own PDP software, but also share it instantaneously with your colleagues around the world. For example, when you attend conferences in future, instead of sitting there half asleep and forgetting most of the pearls of wisdom which you hear, you will find yourself and everyone around you behaving quite differently; you will all sit there typing furiously into your PDAs and transmitting the new ideas and information to your colleagues immediately, enabling you and your company to maximise every possible advantage immediately from your attendance at that conference.

If this seems in any way a little far fetched, then it is worth noting that such a system is already being used by police officers in Derbyshire to handle scene of crime statements. Police officers have been issued with Apple Newton handheld computers. There is no keyboard; the officer simply writes the information on a touch-sensitive screen using a special stylus. The officer fills in a special form which appears on the screen, with all the relevant questions which need to be asked. The responses are then transmitted back to the police station over cellular telephone links for instant action.

My colleague Ron Young suggests that technology might progress even further than this. He tells me that voice recognition is definitely a technology which will have a major impact in the future. Furthermore, since we are all increasingly affected by information overload, so-called intelligent agents are being developed which act as filters and sorters of information, automatically sorting out items which the software considers to be important to you on the basis of previous information that it has acquired about your interests. Perhaps in the future, if you are extremely busy, you will not necessarily need to attend a conference yourself; you might instead simply send along your PDA, loaded with voice recognition and intelligent agent software, and ask that it is placed on an empty seat at the front!

8.10 Learning as a two way process

Remember, learning has to be a two way process. Only if you are prepared to go out of your way to share your experience, ideas and information with others will they do the same for you. The old ways of keeping information to yourself in the belief that superior knowledge is power must be ended. Instead, think back over this chapter and identify those things which you can commit yourself to doing to use software to maximise that two way flow of experience between yourself and others. Then write these commitments into your PDP.

Chapter 9
How to make the best use of distance and open learning

'But above all try something'—F. D. Roosevelt

One of the most cost effective approaches to CPD can be by using distance learning. There are many excellent packages available—for example from the Open University—which for a very reasonable price provide an effective mix of printed study material, videos, assessment questions, tutorial support and vacation schools. Indeed, for several years when I headed training at GEC Telecommunications and GPT the company was the single largest user of the Open University, with a few hundred employees each year studying courses at undergraduate, MSc and MBA levels. One of the reasons why such material can be so highly cost effective is that employees usually study the material in their own time, with the company often paying the course and summer schools fees in return, sometimes also providing paid leave for employees to attend the summer schools.

The two terms used in the heading of this chapter are:

(i) *distance learning*—this is where someone is provided with a package of materials, such as from the Open University, and studies at a distance from the course provider.

(ii) *open learning*—open learning can encompass distance learning, but also frequently refers to the provision of facilities such as computers, audio learning facilities and learning libraries. Employees can usually use such facilities in both company and their own time.

It is worth noting that these are my definitions. These two terms are widely used, but ill defined and not even the experts seem to agree on their meanings!

There are, also, many different ways of using both distance and open learning. As with every other aspect of CPD, the context is extremely important. In addition, there is an ongoing convergence between distance and open learning on the one hand, and the approaches using software described in the previous chapter. In a real sense, for example, groupware can be seen as an important form of open learning.

Let us therefore first see how two organisations, which both happen to be in the communications business, approach open and distance learning.

9.1 Opening up learning in GPT

'Anything that you can learn by your own efforts always has a special value,' says Peter Harris, GPT's employee development manager. 'That is why the technique of open learning has become so popular. At GPT people are encouraged to discover things for themselves, because helping them to develop will improve the company's business performance.'

GPT views open learning as a valuable medium for enhancing employee development. This has resulted in a significant investment in dedicated facilities, and there are open learning centres on five of the major sites.

A variety of methods is used to provide instruction. These include:

- computer-based training (CBT)—this is the most popular learning method. The best courses use animated graphics, text, summaries and questions to present information in a clear and logical way;
- interactive video (IV)—a video disk player, controlled and supported by a PC, combines sight, sound and student activity. This is mostly used to develop the interpersonal skills required of managers and supervisors;
- video, audio and text—these are often combined for maximum effect, since they complement each other well. Video is ideal for demonstration, simulation and role play. Audio is the first choice for language training. Text is explanatory and provides a record for future reference;
- practical experiment kits—these usually consist of text supported by equipment designed to allow practical work.

The materials can be studied either standalone, or used in conjunction with other methods of training. A particular example of this is language training where methods used range from intensive, tutor-led courses backed up by video and audio tapes, right through to employees learning entirely from video courses.

Some of the benefits have been:

- meeting training and development needs identified through a personal development programme;
- used as a cost-effective alternative to external training;
- preparation for advanced external training courses;
- revision of skills which are a little rusty;
- supplementing other training.

In addition to on-site facilities, there is a special relationship with the Open University (OU). Over 200 employees study OU courses annually, mainly at post-graduate level. Students study both technical courses and courses supplied by the Open Business School, and they are supported by on-site tutorials and materials specially tailored to the needs of the telecommunications industry.

The cross training of engineers from hardware to software has been one example of the way in which the relationship with the OU has directly and positively affected the development of GPT's business.

9.2 Opening up learning in BT

BT has opened up a new distance learning centre (DLC) as part of its management training centre at Martlesham.

The DLC has 16 interactive video workstations of which four supply language training, four are video booths and two are audio. There is also a room set aside for group learning and discussion.

The wide selection of videos includes over 1000 different titles, which can be viewed in the centre or taken away for use in the office or home.

The facilities include over 40 interactive video subjects, divided into over 70 separate modules; these cover a range of interpersonal skills and languages. In addition, there are over 100 audio cassettes to choose from, computer-based training materials and a large selection of books.

'One particularly helpful feature of the centre's resources is that they come in different media, suitable for people like myself and many members of our team who travel frequently,' says Chris Webbley, departmental quality manager at Martlesham. 'So I have borrowed audio cassettes to make better use of time while driving on the A12, taken video cassettes to view at home and used the more powerful features of interactive video at the centre itself. This has helped me in fulfilling my personal development plan, for example.'

Looking to future developments in learning technology, there are also ten workstations equipped with CD-ROM, enabling employees to use the greatly enhanced power of interactive multimedia.

9.3 Well proven methods

Both distance and open learning are becoming increasingly popular methods for CPD. These methods are now well proven and widely used by staff across a range of industry and commerce.

Some of the advantages to individuals are:

- flexibility of study patterns and location: they can offer learning on demand, equivalent to just in time training;
- less interruption of work and domestic/family arrangements;
- freedom to work at one's own pace, in contrast to, for example, an intensive one-week course;
- they are likely to be a much more effective means of learning if properly organised. For example, you can interleave study of a subject with gaining experience in its application.

Do, however, recognise that the structure of distance learning can vary widely between different course providers. In some cases there are rigid requirements on timescales and subjects to be studied. In other cases, such as with the Open University, programmes are very flexible in terms of the mix of subjects which can be studied and the duration of possible study, with standalone modules which can be used without necessarily intending to obtain any formal qualification.

9.4 Possible delivery media

The distance and open learning materials that you use can consist of one or several of the following media:

- printed paper;
- audio cassettes;
- video cassettes;
- computer-based delivery.

Think carefully about the type of medium which is likely to suit you best. You may, perhaps, prefer printed materials on the grounds that these are likely to be the cheapest and easiest to use anywhere, or possibly you may feel that the high technology involved in using computer-based material will be more fun and stimulating. The more sophisticated the medium, the more expensive it will be to produce, so the more it is likely to cost you or your employer.

Let us look at each of these types of medium in turn.

9.4.1 Printed paper

Printed paper in the form of workbooks is still the most widely used open learning medium, either on its own or in combination with other learning media.

Workbooks usually have spaces for the learner to write notes and responses to the exercises provided. This approach has the advantage of great flexibility. These types of workbooks can be used wherever and whenever convenient. In addition, they give you a permanent record of the learning experience, including your own responses and ideas. There is no need for expensive equipment such as computers or video players to be available to the student.

In general terms, there is much less information on each page of an open learning workbook than on those of a textbook. There will also be more frequent exercises to test the understanding of the student. It is not unusual for three or four hours of study (equivalent to half a day of a training course) to be contained in an A4 size workbook 20 to 30 pages long.

Almost all open learning packages contain a printed element, even where the major medium of learning uses high technology. This printed element has the great advantage that it can be used as the student's personal record of learning, which can be retained. Some open learning packages extend this concept by adding a workfile, specially designed to hold additional written work, drawings, charts, etc., forming part of the exercise work.

9.4.2 Audio cassettes

Audio cassettes can form a very useful supplement to printed workbooks and other media. They are not usually used as the principal method of delivering learning, with the major exception of some language courses.

This medium allows students access to expert views, interviews and discussions in a far more convenient and interesting manner than could be achieved in print. One successful open learning pack on an electronic topic uses audio cassettes to talk students through complex diagrams.

It can be argued that audio cassettes can achieve little which could not be achieved better using video cassettes. Audio cassettes do, however, have one major advantage for the user—they are more convenient to use. For example, they can be used during car journeys and, with the popularity of the Walkman type of personal audio player, can be listened to almost anywhere else.

9.4.3 Video cassettes

Video on VHS cassettes is a popular delivery medium for open learning. It is almost always used with printed workbooks, which students retain as a reminder of their learning experiences.

Coupled with exercises in accompanying workbooks, requiring students to use the ideas being studied, video can provide very efficient and effective learning experiences. It allows students to view a wide selection of events which are very difficult to present in other ways. These can range from viewing dynamic physical processes to observing the use of communications skills in meetings and interviews.

Many successful open learning packages are based on a combination of video and workbooks. These packages vary in the role played by the video element. In some the video is the principal teaching medium and the workbook supports the material on video. At the other extreme, the workbook is the principal teaching medium, with the video adding illustrations and an occasional change of medium. Both these extremes, and a range of mixes between them, have been used successfully. The most important factor is that the videos and workbooks are designed together to offer an integrated package of learning.

9.4.4 Computer-based delivery

Computers provide an extremely effective means of delivering formal learning programmes—as well as providing the many different and exciting CPD opportunities discussed in the previous chapter. For example, they are capable of interacting with a student's responses in a manner only surpassed by a knowledgeable human tutor. They provide an ideal medium for implementing the best approaches to open learning. As one example, a computer-based learning programme not only allows students to study at their own pace; in addition, by continually testing their understanding, and responding accordingly, it can route each student along the most effective learning path.

Modern computers have sufficient power to provide excellent graphics and video-quality animation to enliven and reinforce learning messages. Recent developments associated with data storage on CD-ROM, now holding over 600 megabytes of data, allow soundtracks and video to be incorporated into computer-based programmes to produce so-called multimedia programmes. These multimedia programmes run on a variety of systems such as multimedia PCs, CD-I and IV (interactive video). In essence, they all provide the facilities needed for computer-based learning delivery, but vary in the level and quality of additional features such as video. Although these additional features can greatly enhance the effectiveness of a learning package, it must be noted that the quality of the underlying learning programme design is still the most important factor in a successful package.

The only key disadvantage of computer-based learning delivery is the high initial costs incurred in designing and programming the learning packages. This means that there are relatively few packages aimed at niche markets such as engineering. In addition, there is a need for a suitable computer system for the user. This latter disadvantage is unlikely to apply in the case of most engineers these days, who almost all have access to suitable computers.

The high initial cost has to be measured against the number of eventual users. For a large number of users, the cost per student can become relatively low.

Increasingly, we are likely to see the Internet used to deliver high quality training courses, incorporating high quality graphics, audio and video.

We are also almost certain to see an explosion in the use of Lotus Notes Domino servers to deliver such packages over the Internet—with the advantage to the providers that they will retain control and security over the material at the same time as using the open access strength of the Internet. Linking this combined Lotus Notes/Internet approach with structured access to experts and tutors could in numerous ways revolutionise how many of us approach CPD. A new distance learning package, Lotus Notes Learning Space, was launched by Lotus in 1997 with access available over the World Wide Web as well as Lotus Notes.

9.5 Choosing an open or distance learning course

One of the advantages of using open or distance learning courses is that often the study materials are available for inspection before you need commit yourself to taking the course. You can sample not just the content, but also the style of presentation. In contrast this is next to impossible to do with most conventionally presented courses.

Dr. Mike Lee, a consultant who has considerable experience of producing and applying open and distance learning, and who has been a prime adviser for this chapter, comments:

'Take advantage of the availability of course materials which can be sampled before deciding on a particular one to study. However, be careful when comparing open and conventional courses not to be over critical of the former. By being available for detailed inspection it also makes itself available for criticism. In contrast there may be relatively little information on conventional courses. Remember that we all have a tendency to be optimistic about things of which we know little.

'Open learning, especially when it includes video or multimedia, is expensive to produce, and the glossier the presentation the more it costs. CPD for engineers using open learning is not that large a market, and producers of such material are unlikely to make an acceptable profit by spending huge budgets on products aimed at this market.

'The large markets in this area, in contrast, are often for those qualified to less than graduate level and this is the area which attracts higher budgets and offers glossy materials. Most of these materials are fun to use and offer excellent training to their intended audience, but many graduates find them simplistic and patronising. This may or may not apply to you, so check them out anyway.'

There are open learning packages aimed specifically at engineers, such as the IEE's Management for Engineers series. Courses like these are pitched at graduate level and are more analytical than the more basic packages. They were produced on smaller budgets and use much simpler presentational styles, but they offer excellent learning opportunities and are highly rated by engineers who have used them. When he was head of open learning at the GEC Management College, Dr. Mike Lee set up and monitored courses based on similar packages in GEC companies. Some 90 % of the engineers completing the courses commented favourably on them and said that they would choose similar courses for further CPD activity.

Further advice from Dr. Mike Lee is:

'When choosing open learning, examine packages closely and remember the old adage of not judging a book by its cover. Good open learning design does not necessarily result in lots of glossy pictures and fast action. The most important features are the content, its clarity of presentation and its suitability for your purposes.

'If you are abreast of developments in computer technology, you may be surprised that many so-called multimedia courses drastically under utilise the potential of the medium. This is largely because many of these courses were produced some years ago when PCs were much less powerful than they are now, and the high cost of development deters the producers from producing new versions. As above, judge these courses on their content and their ability to meet your needs.

'Newer multimedia courses make much fuller use of the opportunities offered by current PCs. One example is a recent multimedia programme from the Institute of Physics, designed to help scientists to make the transition from university courses to working in industry and to plan their future CPD; this is based on a database of specially recorded interviews with practising industrial scientists, which users can browse through according to their particular interests.'

9.6 Using open and distance learning

One of the features of open and distance learning is that it will include plenty of guidance on how to use your particular pack or course. So here are some bits of advice on more general ways in which you can maximise the effectiveness of your studies.

The first point is not to regard open or distance learning as being separate from other modes of learning. These different modes are simply different aspects of the totality of learning opportunities open to you and should be treated as such. There is no reason why you should choose to study a particular topic either by open learning or through conventional courses alone. Use open learning with other means of study to give yourself the best mix.

Open learning gives you freedom to arrange your study time as you wish. This can be a two-edged sword. On the one hand, this freedom brings the responsibility for planning your study time such that you study efficiently and effectively. As a graduate engineer, you are unlikely to need much advice on how to plan your study time, but the flexibility of open learning packages tempts users either not to plan in the first place or to wander from the plan at the first minor distraction. Aim to avoid falling into these traps. Use your personal development plan to organise your learning and to capture the key learning points.

You should avoid the loneliness of the long distance learner by discussing and sharing your learning with appropriate other people. If you are studying on a major distance learning course such as one with the Open University, then contact with other students is usually arranged for you by the course providers using contact groups, weekend schools, etc. One distance learning MBA now makes use of the Internet to facilitate contact between geographically separated students. Do not underestimate the importance of these arrangements and use them as much as possible.

If you are studying a smaller open learning package, then you will probably have to set such contact groups up yourself. Identify people who will be interested in what you are doing. These will usually be people whom you know through work, although a partner or friend may be a good choice with topics such as communications skills!

One obvious candidate with whom you should regularly discuss your learning is your manager. She or he should certainly be aware of your training needs and have a vested interest in you acquiring new skills. The sort of help a manager is particularly suited to give you is to put the generic content of the open learning package into the specific context of your job and your organisation. Another obvious person who can fulfil these roles is your mentor, if you have one. Even if you do not need to consult your manager while studying the package (for example, if you have a mentor with whom you are regularly meeting), then it is likely to be very important and highly beneficial to both of you to review your studies with him or her after completing the course.

Another good candidate is someone who has studied the same package already, or perhaps a different package or course on the same subject. Approached in the right way (which is likely to include you letting them know that you do understand that they do not have unlimited time available), most people are flattered to be asked to help and are very generous with their time.

Another possibility is that you may be able to find a partner for your studies. If you know someone in a similar position to you, then they may be interested in studying the same pack in parallel. This works particularly well where your joint training needs arise from a joint work activity where you apply your new skills and knowledge together to complete a common work objective. This just in time training, where you have an immediate application for your new skills, is probably the most effective form of learning.

An extension to the idea of having a study partner is to form a study group of perhaps five or six people with the same study need. All the people in the study group can study the same material and meet regularly to discuss what they have learnt, as well as to carry out group exercises. If a more experienced person is prepared to act as a facilitator, then this works even better. This arrangement has worked successfully in a number of companies and the IEE Management for Engineers series includes instructions on how such groups operate.

9.7 Distance learning and the IEE

Let us look now at the very wide range of distance learning opportunities open to you through one institution, the IEE, as an example. There are also many other institutions offering such opportunities, although the IEE is in fact by far the largest.

The IEE first became involved with distance learning by producing videos of IEE vacation schools (one-week courses run over the summer at various universities, with a high industrial input) in 1988. This developed into a formal publishing activity in 1991.

The Institution now provides a wide range of video and computer-based training packages for engineers' CPD. These come in the form of:

- high quality modular learning materials for self study;
- additional material for tutored courses.

The advantages of these materials are:

- their cost effectiveness;
- their convenience for study as required;
- the flexible way in which engineers can organise their study time.

There is a wide range of study material available, so one action you might therefore build into your PDP is to obtain a copy of the IEE distance learning course portfolio.

Another action you might consider is obtaining a copy of the Institution's catalogue of publications and information services. This will provide you with details of the Institution's range of books, in addition to:

- the conference programme for that year, as well as details of the previous year's published conference proceedings;
- colloquium digests from the previous two years' colloquia;
- information packs and bibliographies based on the INSPEC database (to be explained later in this chapter) and the IEE library collection;
- wiring regulations and associated trade publications;
- Public Affairs Board publications;
- details of the IEE's journals and proceedings;
- details of the INSPEC database of science and engineering abstracts and INSPEC abstracts journals.

9.7.1 INSPEC

INSPEC was formed by the IEE in 1967, based on the 'Science abstracts' service which had been supplied by the Institution since

1898. It is recognised as the leading supplier of services providing references to published information in the fields of physics, electronics and computing.

This is an extremely important potential source of distance learning, in the sense that it enables you to identify sources of reading material in a wide range of subjects. Reading is undoubtedly the single most undervalued and underused source of CPD, and a database such as INSPEC is potentially very valuable.

Although based in the UK, eighty per cent of INSPEC's sales are overseas, emphasising the high respect in which it is held. In 1995 INSPEC merged with the former PHYS database and 'Physics briefs' abstracts journals. As a result there are over 300 000 records added each year to the existing total of over 4.75 million. The source of material for these records is the scientific and technical journals, conferences and other publications produced throughout the world in a wide range of languages. The contents of over 4200 journals and some 2000 published conference proceedings, as well as numerous books, reports and dissertations are regularly scanned by INSPEC staff for relevant articles to abstract and index for the database.

There are three journals which contain short summaries of papers published in the subject areas concerned. These are fairly expensive (of the order of £1000 to £2000), so they might be viable only for large organisations to buy.

The cheaper alternative is to access the INSPEC database online from a number of host systems. Low-cost data communication networks, or the Internet, give most countries easy access to these services. All that is needed to search the database interactively is a password and a terminal which connects the user to the network via the local telephone system. Alternatively, you may use one of the many information brokers providing this service, such as the IEE's technical information unit. Thousands of organisations worldwide use INSPEC online to aid their research and development activities, so why not persuade your organisation to do the same?

Many organisations run their own computer-based information services using databases produced in house or purchased from external sources, including from INSPEC tape services.

In addition, INSPEC is available on CD-ROM.

How about checking out, therefore, whether your organisation has any of these types of information database available. If so, then try to make regular use of them. If not, then ask whether your organisation should be persuaded to obtain such databases.

9.7.2 'Electronics letters'

'Electronics letters', published 25 times a year, is a prestigious publication recognised as the leading journal for the fast publication of short papers in letter form describing research across the whole spectrum of electronics.

This is also available as an online electronic journal distributed to subscribers via Internet and dial-up telecommunications networks.

9.8 The Open University

Mention must also be made of the excellent range of courses available from the Open University. These include:

- an MSc in computing;
- an MSc in manufacturing;
- an MBA in technology.

The first two consist of courses which may be taken as free standing units or combined to give a postgraduate diploma which may be converted to an MSc by completing a project and dissertation. The MBA is a mix of compulsory and optional courses.

9.9 Reading—the most underutilised source of CPD

As stated earlier, it is almost certainly the case that you can make much greater use of reading to improve your CPD.

Therefore, take some time now to put together a list of the books and other publications which you have read in the past year or two. Make this list a part of your PDP. Make a note of the ideas, information, etc., which each of these books provided for you.

Next, take some time to list the books, journals and so on which you should be making time to read over the next three months. Make a note in your PDP of this list and identify:

- where you will obtain these—will you buy them or borrow them from your local or institution library, for example?
- when you will be able to read them—in the evenings, at weekends, on the train, on aircraft journeys, or when?

Again, enter all this information in your PDP.

Keep asking yourself questions such as these. Do you, for example, regularly read your institution journals and keep a record in your PDP of what you have learnt from them? How about reading newspapers such as 'The Financial Times', or journals such as 'The Harvard business review'?

9.10 And how about writing?

So far we have looked in this chapter at many ways in which you can access information through distance and open learning.

Now ask yourself how much you are contributing yourself to the world's sum total of knowledge. Have you ever written any articles, for example for your company journal, newspaper or newsletter, or for your institution's journals?

They also say that 'there is at least one book in each of us'. Would you like to write a book at some time? If so, what stops you? You will probably find that most of the barriers are in your mind—and if you are sufficiently determined these can be reasonably readily overcome.

So how about finishing this chapter by making a list of the articles and books you have already written, together with those you target for yourself in the future—and of course then putting all this information into your PDP?

Chapter 10
The role of the universities and other providers

'If you think education is expensive, try ignorance'—Derek Bok, president, Harvard University

It will hopefully have become abundantly clear to you by now that my strong belief is that CPD can only be effective where there is an appropriate structure and context.

All too often CPD is seen to be mainly about courses, and universities and other providers need to be very wary of the dangers of failing to integrate their courses, consultancy and other undoubted expertise with the real ongoing needs of their clients.

One of the best examples of how to do this is in Motorola, a company which believes very strongly in a coaching style of management and in identifying the key competencies for each employee.

10.1 Motorola University—a catalyst for change through continuous learning

Motorola, covering a wide range of products from semiconductors to mobile telephones, is a highly successful company, with more than 140 000 employees worldwide and annual revenue expanding at over 25 per cent between 1993 and 1995. Part of the reason for this success has been the company's adherence to every employee participating in a minimum of 40 hours of off the job training each year, in addition to ongoing on the job learning.

A study of Motorola by an outside company shows just how effective training can be in those plants where the context is right. Where employees absorbed the relevant quality tools, as well as the associated process skills, and where senior managers reinforced the learning with appropriate coaching, there was a $30 return for every dollar invested in training, including the cost of wages paid while employees were being trained.

Continuous learning linked to raising quality levels

Motorola is committed to providing continuous learning for every employee. This has been linked to reducing the time taken to design a new product from between three and seven years down to 18 months. Even more importantly it is linked to the objective of achieving six sigma standards of quality throughout the company.

Six sigma refers to six standard deviations from a statistical performance average, equivalent to just 3.4 defects per million. (In comparison, airlines achieve 6.5 sigma in safety—counting fatalities as defects—but only 3.5 to 4 sigma in baggage handling. Doctors and pharmacists achieve an accuracy of just under five sigma in writing and filling prescriptions.)

Motorola aims to achieve a ten-fold improvement in quality every two years, with total customer satisfaction as the key aim.

The importance of teamwork

Tonnes Funch, director of Motorola University, says: 'Motorola is committed to lifelong development. Learning needs to be stimulating and we involve members of employees' families if appropriate. Most importantly, our staff need to know how to work together effectively and they have people to help coach them. We place a lot of emphasis on facilitation, counselling and coaching skills. Generally, we have no supervisors and may have 300 employees reporting into one manager. There are 53 000 people working in 'total customer satisfaction' (TCS) teams around the world. These TCS teams are cross functional, problem solving teams, which often include customers and/or suppliers, and they have learnt how to use each other's competencies to solve problems. They are shown how to be open, tolerant and how to listen. Creativity and innovation are key and employees set themselves very tough targets—often ones which they think are impossible, and yet they do achieve them!'

The Motorola University concept

Motorola is committed to being best in class by continually searching for ways to do things better with an efficient and skilled workforce second to none in the world. Motorola University is the catalyst for making this happen by encouraging and supporting a shared individual and corporate commitment to continuous learning.

Motorola's worldwide education and training community is a federation—a gathering together of all Motorola training organisations called Motorola University. Through collaborative joint venturing, the partners in this federation develop the experience and expertise to enhance corporate competence by:

- providing visibility to major skill/job shifts and emerging markets;
- leveraging educational assets—human, technical and physical—and sharing resources;
- improving learning efficiency, eliminating redundancy and reducing training costs;
- reducing the cycle time of training needs assessment and course development;
- transferring knowledge through common training platforms.

Auxiliary resources for competency building

Supporting the learning process across Motorola are:

(i) *The National Technological University (NTU)*—this instructional satellite network in the USA delivers courses, provided by leading universities and other organisations, to Motorola employees doing masters degrees and other professional studies.

(ii) *Motorola University Press*—the knowledge and expertise of Motorola employees, as well as the culture, history and key beliefs of Motorola, are disseminated via the printed word throughout the corporation and to customers and suppliers by this publishing operation.

(iii) *Educational and corporate partnerships*—partnerships have been formed with institutions of higher education worldwide, for example with Edinburgh, Warwick, Cambridge and Erasmus Universities, the Judge Institute and Compiègne University of Technology in Europe. Motorola University certifies lecturers from these many institutions who then deliver courses on the company's behalf. Motorola University also provides training to major suppliers as well as customers in order to maintain the best supply chain.

In summary, therefore, this is an excellent case study showing how effective CPD providers can be, both by working in close partnership with employers and in the right context.

10.2 A partnership—the Warwick Manufacturing Group

One university which has built up a considerable reputation internationally for integrating its expertise to meet the specific needs of industry is Warwick University.

In 1980 Professor Kumar Battacharyya launched his vision at the University of a new national centre in the UK for the development of top-rank engineers and engineering managers. To achieve this vision he has pioneered many innovative teaching and research methods—with industrial partnership replacing the traditional academic approach.

His flagship is the Integrated Graduate Development Scheme (IGDS). This masters degree programme, which has now been adopted in a number of other universities, is based on a joint definition of all aspects by both industrialists and academics. Half of the lectures are provided by practitioners, mainly in industry, with just the core fifty per cent coming from Warwick. Not only is there immense flexibility in developing new material, but another pioneering step was to make the degree part time, modular and assessed entirely through in-company assignments and project work.

There are now more than 400 participants sponsored each year by over 40 manufacturing and service companies. These include Rolls-Royce, British Airways, British Aerospace, GKN, London Electricity, Lucas, Raychem, Rover Group and Short Brothers. Awards are available in engineering business management, design systems, manufacturing systems or information technology—with all participants tailoring an individual programme from 50 possible modules.

This has been complemented since the mid 1980s by the Integrated Manager Development Scheme (IMDS), a similar part time modular diploma programme which is designed for middle managers, mostly those without formal qualifications. The success has been such that there are now as many participants on this scheme as on the IGDS.

Both schemes have now been adopted in Hong Kong, with other activities growing rapidly throughout the Pacific Rim. These schemes are also complemented by short awareness programmes for company board members and other senior executives, as well as a strong applied research programme—in all cases collaborative with industry.

10.3 CPD—providing a flexible workforce

A rather different example is Wearside College, which initially developed strong links with one particular employer, Nissan. These links have then been considerably expanded to help other employers in the region.

In March 1984 Nissan Motor Manufacturing (UK) Ltd decided to locate its first European car manufacturing plant in Sunderland.

Establishing motor manufacture in a region with no such previous experience demanded a radical approach to the skill and training requirements of the workforce. As a company dedicated to quality and flexibility, Nissan recognised that technological change would place increasing demands upon the skills and knowledge of their maintenance personnel, who needed to be multiskilled. They had to be able to diagnose problems quickly in systems combining mechanical, electrical, electronic and software technology and rectify the fault. Wearside College was approached by Nissan to develop a multiskilled technical training progamme to meet this new challenge.

Working together, using an analysis of the company's skills profile, both Nissan and the college recognised that traditional methods of training in single disciplines would not produce the required skills. As a result a 60:40 broad division between electrical/electronic and mechanical/production was used in the design of the training programme. Other essential elements were that it should lead to nationally recognised qualifications, that it should combine practical and theoretical elements and that it should form the basis of a CPD programme for the trainees.

Continuous development programmes are provided for everyone at Nissan. Technicians completing the multiskilled training programme have the opportunity to continue their training in higher education. Open learning facilities are extensive and available to all employees, with a wide range of learning and training materials.

The success of this major training initiative is such that 24 companies have now identified similar multiskilled training needs and formed a local consortium with Nissan. Their diverse activities range from offal processing to adhesive manufacture. This common approach to a multiskilled workforce is contributing to the growth of the region and helping to attract inward investment.

A number of schemes are now available on a modular basis from consortia of universities, employers and training providers. Let us next look at two such examples.

10.4 The Continuing Professional Development Award Scheme

A consortium of higher education institutes (HEIs), companies and private training providers has developed the Continuing Professional Development Award Scheme. This has been in existence for over six years and is based around quality assurance of courses/programmes and of the HEIs and organisations providing them.

The scheme aims to meet the four key needs shown in Figure 10.1.

A tariff of 120 hours of CPD, which is gained by accumulating credit from courses within the scheme, has been set as the point at which an award is given. For convenience the tariff is divided into six credits, each of 20 hours, so that even modules as short as ten hours (a half credit) can be counted. Courses can be in any subject area chosen by the individual.

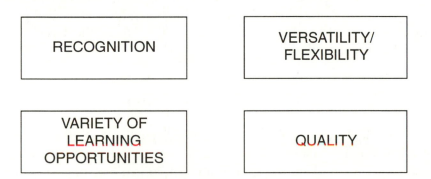

Figure 10.1   The four key needs targeted by the Continuing Professional Development Award Scheme

For example; a civil engineer might take a wide range of subjects to update herself on current issues and develop management related skills. She could be involved in a contract in Japan. To meet all these needs, she might decide to take the following modules at a number of universities and training providers:

● management of contracts and projects (two credits);
● civil engineering law and contract procedures (two credits);
● introduction to Japanese (one credit);
● finance for nonfinancial managers (one credit).

Over 600 people have already participated in the scheme and it was expected that the numbers would have increased by a factor of ten by 1996.

10.5 TEAM is launched

The Technology Management Programme (TEAM) was launched in July 1993. Students may study just one of the 56 modules so far accredited by the 20 participating universities or take part in the TEAM masters degree programme—which requires the study of a minimum of ten modules, plus a major work-based project.

Each module is equivalent to 80 hours of work and usually consists of 15 hours of precourse work, a 40 hour one week residential course and 25 hours of post-course work which includes a formal assessment and usually contains a work-based assignment. Some modules are also available in distance learning format.

The programme is managed centrally by the JUPITER Consortium which operates a helpline service.

Other effective approaches to CPD can be seen where a provider concentrates on one particular specialism, such as in electrical power engineering and software engineering. These form the basis of our next three case studies.

10.6 Powering CPD—a modular solution

The electrical supply industry operates the largest and most complex of all man-made systems. It is a major user of modern technologies including digital electronics, online plant controllers, expert systems, optical fibre communications and data acquisition and control systems. It supports the investigation and exploitation of novel and environmentally friendly sources of generation, flexible and improved forms of transmission, economic modes of distribution and efficient methods of utilisation.

Unfortunately, many universities have abandoned electrical power engineering in favour of electronics and associated courses, based on the point of view that information technology is the future and power is the past. As a result there is now a shortage of well qualified and experienced electrical power graduates in the UK.

The Manchester Centre for Electrical Energy (MCEE) at the University of Manchester Institute of Science and Technology is now providing a solution—a uniquely structured postgraduate programme. With a history of providing such courses for industry stretching back as far as 1957, and an MSc in power engineering since 1962, MCEE has now launched a new modular version. The one and two week modules are each self contained and involve intensive teaching, tutorial and laboratory work. Modules can be studied individually as part of an engineer's mid-career training or can be accumulated towards an MSc, either full time or part time.

The modules are mainly taught by the 16 members of the MCEE, but specialist lectures are given by senior engineers from the electricity supply industry, manufacturing industry, consultancies and from other universities.

Formal professional institution approval of this scheme has been given within the new CPD frameworks such as that of the IEE.

10.7 IGDS schemes hit quarter century—and arrive in Oxford

The Engineering and Technology Programme within the newly formed Engineering and Physical Sciences Research Council (EPSRC), one of the successor bodies to the Science and Engineering Council (SERC), offers a number of schemes which lead the way in industrially relevant postgraduate training.

The Integrated Graduate Development Scheme (IGDS), funded by EPSRC, has been operating since 1979 (see the earlier case study in this chapter about Warwick University) and there are currently 25 programmes in fields as diverse as manufacturing systems, packaging technology, aerospace design, environmental engineering and materials technology.

One of the latest has been initiated by the University of Oxford's Department for Continuing Education, with some 20 students starting in 1993 from IBM's development laboratory at Hursley Park. This particular scheme builds on many years of collaboration between IBM and Oxford University's computing laboratory, which provides staff as tutors under the leadership of Dr. Jim Woodcock. The programme leads to a postgraduate diploma in software engineering with an optional MSc, and has attracted three types of student:

- those early on in their careers, with at least two years postgraduate experience looking for a more specialist qualification;
- some who are much more experienced, but want their knowledge updating;
- a number who were programmers some two decades ago, have since moved into other careers and now wish to move back into modern software development.

The course is based on the rigorous use of mathematical principles to produce high quality software efficiently, and requires a certain level of mathematical competence on entry.

To help prospective students who are perhaps not so confident of their maths ability, the scheme also includes a pre-course programme of individually tailored study units aimed at developing these underpinning mathematical skills. This has proved very popular and effective.

Five of the six one week modules are run at the IBM Education Centre, with a sixth residential module at Oxford. Some ten hours of self study are required before each module, with an assignment taking about 25 hours in the two weeks following. A portfolio of short reports must also be prepared by each student identifying at least three areas where what has been learnt has been, or could be, applied. All this is reinforced by a project based on work undertaken in the company, mainly during the last five months of the course, and presented as a substantial dissertation leading to the award of a diploma, which can be gained over either one or two years.

There are now two further developments. First, with funding from the EPSRC, the scheme has been extended to other companies engaged in software development as from June 1994; these modules are all taught at Oxford, but otherwise the scheme is the same. Secondly, those students who have completed the diploma now have the opportunity to study a further four modules leading to the award of an MSc.

There is one final satisfying aspect of the programme; Dr. John Axford, who helped to initiate the scheme when heading education at IBM Hursley, has developed his own career by moving to Oxford University's Department for Continuing Education and continues his involvement with the IGDS programme.

10.8 Mechatronics or bust

It is estimated that eighty per cent of the workforce required in the UK by the year 2000 is currently employed and that only twenty per cent of the technology which will have been developed by then is now currently available. The gap analysis is alarming, particularly where we review the present capability of our wealth generating industries to attain world manufacturing standards.

The UK's industrial workforce has been reduced over the past 25 years by forty three per cent because of:

- loss of competitive edge;
- lack of reinvestment of profit into updating resources by British industry.

The overall picture of the UK industrial scene is one that is still in decline and will remain so unless there is a step change in the attitudes towards education and training by educational institutions and industry, and a planned reinvestment programme for re-equipping with state of the art technology. The education, training and retraining programmes that must be embarked upon in the UK must relate to the product needs of future world markets. There is every evidence that these will be of increasing sophistication and intelligence. Manufacturing companies will require interdisciplinary engineers and technicians to undertake the R&D, manufacture and maintenance of these products.

With the availability of micro and nano machining techniques, and the capability to design in virtual space, products will consist of complete systems. This will require a workforce trained across disciplines and in the application of the mechatronic philosophy.

Gwent College of Higher Education, in collaboration with Innovative Technologies in Education (ITE) Ltd, has set up the Gwent Mechatronics Development and Training Centre in the UK, which is one of only six in the world. The CIM centre is part of the total mechatronics training facilities within the Faculty of Technology. Accredited training in Siemens PLCs and Autocad, electro-pneumatics/hydraulics, communications and control supports the total education and training programmes available in mechatronics.

Engineers and technicians would be well advised to include, within their individual CPD programmes, training in the mechatronics philosophy and application. If we do not create a workforce that is interdisciplinary in nature then we will not create the industries able to compete in tomorrow's world markets.

There is now a need for a quantum change in attitudes towards training by industrialists and the provision of mechatronic-type education programmes if we are to regain our competitive edge. There is very little evidence in the UK that this message has been received or understood.

One of the interests that I have in this case study is that the word mechatronics was invented in 1964 by Professor Takashi Kenjo, my Japanese co-author. Takashi has had many books on electric motors and circuits published in many languages. However, for his first book published in English by Oxford University Press in 1984 he was told that there was no such word as mechatronics in the English language and had the word refused. These days there are even professors of mechatronics in universities!

Another important aspect of technology is in the medical field. The next case study shows how one career move can be into rehabilitation engineering.

10.9 Rehabilitation engineering through CPD

Rehabilitation engineering is a term used for the application of engineering principles to the alleviation, as opposed to the treatment, of disability. Examples include the development of special supportive seating for people who have to live their lives in chairs, mobility aids (e.g. wheelchairs), artificial limbs, communication aids, environmental controls and other aids for daily living.

Rehabilitation engineers need a wide range of skills and experience beyond their basic engineering training. First, they should know about the medical conditions of their clients so that the aids which they provide are appropriate. Secondly, they should be sympathetic and effective communicators, both for understanding and training clients and for contributing to multidisciplinary consultation with their clinical colleagues. Thirdly, they should be capable managers because, with the formation of NHS trusts, their role now includes the setting up and control of contracts with external suppliers and neighbouring health districts.

The Centre of Rehabilitation Engineering, CoRE, is part of King's College, London, and is funded by the Department of Health to provide educational underpinning for engineers in the NHS. CoRE runs part-time courses leading to a college certificate and a diploma. These recognise the standards laid down in the basic and advanced training manuals in rehabilitation engineering issued by the Department of Health and supervised by the Biological Engineering Society.

'After many years of steady development, technology in this field has started to change rapidly, with a lot of new equipment becoming available,' says Alan Turner-Smith, senior lecturer at the centre. 'Pressure of work in the Health Service means staff cannot afford to take long periods away to adjust to the new standards of practice. So CPD, available in short courses such as ours, is a highly suitable way of introducing new skills. Many of our students have served in the National Health Service for a number of years, after qualifying with an ONC, HNC or HND, and are now keen to raise their professional standards to match the new demands in modern healthcare. Other students, including a number with degrees, are converting from other engineering occupations. Whatever their background, we can identify those who take up CPD as the people who take their jobs seriously.

'The certificate course at CoRE involves three one-week residential units and a one-week clinical placement,' continues Alan Turner-Smith. 'In between, students complete worksheets to follow up study themes, write up a library assignment and report on a project at their workplace. A new CAT points scheme is being prepared for 1996, which will add flexibility to CoRE's CPD programmes. This will also link with the rehabilitation engineering option in the University's MSc in medical engineering and physics. We are also exploring the development of NVQs in rehabilitation engineering. With ever more technical changes on the horizon, CoRE has its own research programme and has developed strong links with rehabilitation technology training programmes in other European countries.'

In addition to running these courses and two annual national conferences for rehabilitation engineers, Alan edits 'REview', the Centre's newsletter, and has reprinted several articles from 'CPD link', The Engineering Council's CPD newsletter. He is a member of the Biological Engineering Society, which has a CPD scheme that is to be closely modelled on that of the IEE. Alan is himself a physics graduate who did research into laser physics at Oxford before joining Philips Research at Redhill. He then developed a number of medical engineering technologies at the Oxford Orthopaedic Centre prior to moving to his present job in 1992.

10.10 Building on law

Another specialist area in which engineers might need expertise is that of law, and one university offering such expertise is Strathclyde.

Strathclyde University Law School offers an evening based postgraduate course in construction law, which started in January 1995. This is specially designed for professionals working within the construction industry—as well as practising lawyers.

Diploma students follow the same four module programme as LLM Law students, except that there is no requirement for a dissertation.

Each standalone module has one three-hour class per week from 5 p.m. to 8 p.m. over ten weeks, and currently one module is taken per semester (although it is possible to take two). Assessment is by examination at the completion of each module.

This course offers useful CPD to construction professionals as well as an advanced qualification from a well-respected Law School.

10.11 Forensic engineering at Glasgow University

A postgraduate course at The University of Glasgow introduced, in September 1995, an imaginative new addition to its CPD training programme. The innovation of a forensic engineering course created considerable interest from within the UK, America and Europe.

This pioneering course, which appears to be the first of its type in the world, was devised by Dr. Colin Goodchild and Dr. John Anderson in the Faculty of Engineering. They believe that a more structured approach is necessary for equipping engineers to deal with the steady increase in legally-constituted enquiries concerning technological failure. Expert testimony across the spectrum of engineering is central to the definition of the cause of failure, the conduct of the scientific investigation and the presentation of results in lay terms.

Forensic engineering is thus defined as the investigation, collection and analysis of evidence of technological failure, with the expectation of presenting this evidence in a court of law. In developing the course it became clear that the topic also attracted interest from the medical, legal, insurance and loss adjustment professions, as well as from the fire, accident and emergency organisations. Worldwide interest also led to broadening of the legal content to include aspects of European law, reflecting the cross-national aspect of the topic.

The original five day course had twenty delegates, including a lawyer from Switzerland. It started with structures and procedures in courts of law, the investigation, collection and presentation of evidence, health and safety, product liability, and engineering case studies across the spectrum of engineering. It continued with fatal accident inquiries, insurance matters, fire, accident and emergency case studies and further engineering case studies. The first three days then finished with forensic techniques in photogrammetry, metrology and computer analysis, quality assurance and forensic engineering in the nuclear industry.

Legal considerations, as well as dealing with press and TV interviews, were covered in the final two days. Legal instruction in the roles of the lawyer, expert witness, expert testimony and judge culminated in a practical courtroom workshop dealing with expert witness testimony in front of a judge.

This course is now expected to continue on a regular basis.

Another specialist area of expertise for engineers with legal implications is in the water industry, which forms the basis of the next case study.

10.12 CPD—insuring against disaster

A recent report in the 'New civil engineer' tells of 'disastrous legal consequences' and 'punitive insurance costs' for civil engineers adopting a blasé approach towards the issue of CPD. The report concludes that an alarming degree of ignorance exists over the Institution of Civil Engineers' requirements on CPD, despite details being contained in members' annual subscription notices. Insurance firms have subsequently confirmed that personal indemnity cover for engineers will be more expensive in cases where CPD records are not up to date. A planned training and assessment programme is the most appropriate method of avoiding problems with CPD records—such as that offered by WTi (originally called Water Training International).

Susan Joyce, engineering product manager at WTi's Millis House training centre near Derby, comments: 'CPD benefits individuals and employers alike. Staff need to keep their technical knowledge up to date, as well as adding to their skill base. Engineering institutions are now beginning to insist on CPD records being kept by members; some organisations are using their quality assurance systems to record that staff have received appropriate training for the tasks which they undertake. This applies to all staff regardless of profession or rank.'

CPD is now becoming obligatory for members of many of the engineering institutions. The number of hours or days varies from one institution to another, but 20 to 30 hours per year is not uncommon. To meet this need, WTi has developed a suite of courses, including one-day specialist events and longer wide-ranging courses. All WTi's engineering courses are designed to meet the needs of staff working in engineering design and management. To keep training events relevant, realistic and up to date, the company works with a large number of practising engineers and specialists in the design and running of courses. It also works very closely with research institutions to ensure that new information, skills and techniques emanating from the latest research are reflected in the content of courses. To complement this training, WTi has invested in the most state of the art facilities, including real-time computer access to industry which allows valuable simulation opportunities.

The engineering courses offered cover a wide range of topics— forms of contract, project and contract management, health and safety, environmental topics, law for nonspecialists, design of water supply and sewerage systems, personal development and civil, mechanical and electrical topics. The presenters are a mix of experts from various sectors of industry. Training exercises form an important basis for all courses and include the design of a pumping station, a contractual dispute or a management problem, all based on actual case material.

CPD is aimed at ensuring that high standards are maintained throughout the engineering professions. Maintaining records is easier if a planned, structured course of training is organised—either as an individual or as part of a team. WTi offers the organisational and technical expertise to ensure that training is provided at the appropriate time, and is recorded and registered through the appropriate channels.

10.13 Eutech—a long and distinguished heritage in CPD

The final case study looks at Eutech, which is part of the multinational ICI and has adopted a very interesting approach to providing its expertise on a commercial basis to the world outside ICI.

Eutech Engineering Solutions Limited has been established as a wholly owned subsidiary of ICI, with the aim of marketing the company's considerable engineering experience and expertise. 250 professional technical staff have effectively moved from being an integral part of one of the UK's foremost engineering companies to being part of a relatively small, but highly expert, company marketing its expertise. This is an interesting example of the way in which industry is restructuring, with implications for both companies and individuals.

The human resource at Eutech is tangible evidence of the success of ICI's CPD programme over many years. Expertise covers a wide range of businesses operating in the process, manufacturing and engineering industries. Staff can provide a comprehensive problem solving capability across all engineering disciplines.

Roy Sallabank, business manager of Eutech's training operation says: 'Eutech has embraced the ICI philosophy of continuous investment in the development of people. Our staff are practising engineers in whom ICI has already invested substantial time, money and resources; Eutech is maintaining that commitment with comprehensive and ongoing training programmes. In particular, we are developing the specialist skills of our engineers, focusing on those areas which are relevant to the needs of our business and which address the major issues and challenges facing the industry as a whole. Through Eutech, companies outside ICI can now reap the benefit of this investment.'

Eutech courses are designed and structured to meet the varying needs of engineers at different career stages. These include developing professional competence, maintaining technical standards and specialist skills. Courses cover a wide number of areas, including:

- safety, health and environment;
- project management;
- mechanical and materials engineering;
- control and electrical engineering;
- civil engineering and construction management.

Eutech can supply consultants or training packages which have been customised to suit specific needs. 'This can make a major contribution to solving the client's problems and delivering the skills needed by their business, both now and in the future, by developing the competence of their engineering resource,' says Roy Sallabank.

Eutech is therefore now benefiting from the investment in CPD by ICI over many years.

10.14 Summary

We have looked at a wide range of case studies in this chapter, showing the extent of opportunities available to most people as part of their CPD.

If there is anything more that CPD providers, employers, institutions and individuals can do to make all this CPD provision even more effective it will be by:

- introducing simple, but useful, competence frameworks for every engineer;

- ensuring that all organisations really do adopt a coaching style of management to fully support the learning process;
- making as wide a use of mentors as possible;
- encouraging all individuals to use some form of personal development plan, ideally using software to directly link learning experiences to each individual's competence development programme;
- building in some form of ongoing feedback to ensure that all such formal learning is maximised as far as possible, as well as fully integrated into each person's ongoing informal day to day learning.

Chapter 11
The role of the professional institutions

'When you can teach someone else what you have learned, then you really know that you know!'

The professional institutions in the UK have been formed over the last century and a half to promote the learning and professionalism of their members. This implies that those who are more expert in a particular subject will pass on their expertise to others less expert— and that in turn these greater experts will benefit by structuring their knowledge in such a way that it can be passed on to others in the most effective way possible. The best way, after all, of checking that you really do have a sound grasp of a subject is to make a presentation or write a paper on it!

The Institution of Electrical Engineers (IEE), the largest of the UK based professional institutions, was formed in 1871 and now has some 140 000 members in almost every country around the world. It really is an international institution. Its first president, for example, was Sir William Siemens, a German who came to the UK in 1843 and founded an electrical engineering company, called Siemens at the time, which eventually became part of Associated Electrical Industries (and where I did my student apprenticeship) and was then taken over in the mid-1960s by GEC. Sir William Siemens played a key role in developing the first electric street lighting in the world, as well as the first transatlantic cable. He married a British wife and became a very distinguished naturalized British citizen in 1859. One of his brothers founded the present-day Siemens company in Germany, and another brother founded a now defunct Siemens company in Russia. Interestingly, until after the turn of the century the Siemens business in Britain was larger than that in Germany. When in August 1995 Siemens announced a £1 billion investment in a new microchip factory in the UK, they naturally chose to make the public announcement in the Siemens Room at the IEE!

One of my responsibilities in the IEE is as the link co-ordinator between the IEE's South East Midland Centre in the UK and the IEE's centre in development in Belarus. I have also chaired IEE training accreditation visits to companies in Hong Kong and Dubai, as well as in the UK. I can therefore testify personally as to how institution activities can greatly benefit their members through enhancing mutual learning internationally.

Paragraph 4 of the IEE's Royal Charter, for example, says:

'The objects and purpose for which The Institution of Electrical Engineers is hereby constituted are to promote the general advancement of electrical science and engineering and their applications and to facilitate the exchange of information and ideas on these subjects amongst the members of the Institution and otherwise for that purpose.'

There are over 40 institutions with their headquarters in the UK, some even older than the IEE (the Institution of Civil Engineers was founded in 1818, for example, and the Institution of Mechanical Engineers in 1847), and numerous equivalents in other countries. In general, the roles of all these institutions are similar at heart to the objects and purpose of the IEE described above, although the UK based institutions and those with similar British ancestry in other countries also have one other key role: that of determining who is a fully qualified and experienced technician, incorporated engineer or professional engineer, allowing the individual to write Eng Tech, IEng or CEng after his or her name. In contrast, in most countries around the world, you are a fully recognised technician or engineer by virtue of your highest academic qualification, albeit that in most cases these countries have rather longer degree courses than in the UK and many of them equate to UK masters degree level.

Much argument has taken place over the years on equivalence between technicians and engineers in these different countries. Across Europe the engineering profession is represented by the Fédération Européene d'Associations Nationales d'Ingenieurs (FEANI), which maintains a register of european engineers who are entitled to put the prefix Eur Ing before their name.

In the UK the umbrella body for the technician, incorporated engineer and chartered engineers is The Engineering Council. This was formed in 1981 on the recommendation of the Finniston Report[13] on the engineering profession, superseding the Council of Engineering Institutions which had existed since the 1960s. The Engineering Council is an independent body authorised by Royal

Charter to advance education in, and promote the science and practice of, engineering for the public benefit.

To be more accurate, the Finniston Report recommended the establishment of an engineering authority as an engine of change, with statutory registration of engineers, but the government of the day considered this too centralist and thus The Engineering Council was born. Its chairman in the 1990s, Sir John Fairclough, initiated a debate on how it should be reformed, to answer many criticisms of its lack of effectiveness in raising the public standing of the engineering profession, and a new organisation to replace it, also named The Engineering Council, was formed in 1996.

Perhaps it would have been much more satisfactory if the Government had indeed accepted the Finniston Report proposal for an engineering authority. This would have included a statutory register of those qualified and the specification and enforcement of a code of practice to be observed by registered engineers. Furthermore, Finniston recommended that 'The Authority should draw up a code of practice based upon engineers' technical competence and continuing fitness to practise, to which registered engineers would be required to commit themselves and breach of which would render them liable to deregistration'.

This code of practice, said Finniston, should require the following of registered engineers:

(a) not to undertake work for which they could not validly claim competence;
(b) to maintain and develop their competence through participation in continuing formation programmes;
(c) to maintain their knowledge of, and to observe wherever appropriate, technical standards, codes of technical practice, health and safety regulations, and other such requirements;
(d) to participate in and to encourage the formation and professional development of other engineers.

It is interesting that it in fact took another decade and a half for The Engineering Council to produce a code of practice for individual engineers, developed jointly with the engineering institutions and which is included, for example, in the IEE's CPD handbook. This states: 'To gain maximum benefit from continuing professional development it is recommended that individual members adopt the following code of practice:

1. Prepare a CPD plan. This should include the following:
   (a) the knowledge, skills, understanding and attitudes to be

acquired or developed taking into account the employer's business objectives, career intentions, short and long term, relevant personal interests and the requirements of the Institution;

(b) the actions to be taken, with responsibilities and timescale, to meet the identified needs.

The plan should be developed, where possible, in conjunction with the employer. Account should be taken of the guidance provided by the Institution for the preparation of CPD plans and the plan should be reviewed regularly (at least annually).

2. Record CPD activities and achievements so that progress towards implementing the plan, and maintaining professional competence, can be demonstrated. The benefits should be evaluated.

3. Provide the Institution annually with a completed record form of the CPD that has been undertaken so that this may be formally recorded and credited.

4. Support the CPD of other staff, and encourage the employer to support CPD as an integral part of professional life.'

In summary, Chris Senior, senior executive, CPD, at The Engineering Council, describes the need for CPD to emphasise:

- the responsibility of individual engineers for continuous improvement and development to ensure competence as professionals throughout their careers;
- the need for development to include a range of technical, commercial, financial and management subjects;
- the use of a wide range of structured job-related activities, including courses, distance learning, in-company programmes and professional institution activities.

While it may indeed have taken The Engineering Council some 15 years to implement the Finniston proposal of a code of practice, there were many important steps taken during that period, and perhaps a code of practice only makes sense once many of these initiatives are in place. For example, The Engineering Council, and Chris Senior in particular, together with his predecessor Bernard Dawkins, was a pioneer in introducing the concept of personal development plans in the UK. It also did much work to organise conferences and provide reading material to clarify what CPD is all about and what might be done to improve its quality.

11.1 A four way partnership

The Engineering Council sees CPD for engineers as being a four way partnership, as shown diagrammatically in Figure 11.1. This partnership between individuals, their employers, the professional institutions and the providers of formal education and training courses is an important one. It should be very clear in every employee's mind, and each individual should be given every possible support by the other three partners.

This chapter is specifically about the role of the professional institutions, so let us look at the IEE as an example of the very wide range of professional activities which a professional institution can provide. These are shown in Figure 11.2; how about taking a bit of time now to make a note of how many of these you take part in and benefit from? Which others might be of value to you? When will you do something specific to enhance the benefit that you can get from your institution? Write down these conclusions and share your thoughts and commitment with someone else—perhaps your mentor, your coach, your manager, or a friend, spouse or partner.

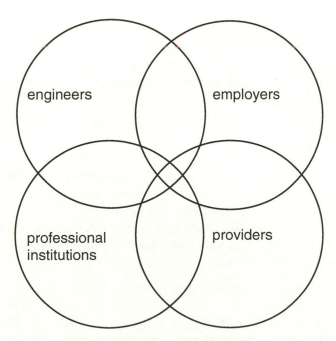

Figure 11.1   *The four way partnership in CPD*

Some of the IEE's many activities:

visits to company plants and sites
conferences
journals
discussion meetings
colloquia
seminars
vacation schools
newsletters
career guidance
professional development record
INSPEC database
books
distance learning

Figure 11.2   Range of activities which a professional institution can provide

11.1.1 The IEE CPD scheme

Increasingly, professional institutions around the world are introducing CPD schemes for their members. Generally these fall into one of three categories:

1 *Voluntary*—the institution member is free to choose whether or not to participate.
2 *Obligatory*—members are obliged to update their competencies through regular CPD by an ethical requirement in the institution's code of conduct. Alternatively, several institutions make CPD obligatory by a byelaw, rule or regulation.
3 *Compulsory/mandatory*—institution members will be disbarred if they do not show that they have had a minimum amount of CPD within a specified period.

The IEE was the first institution, in 1979, to introduce a system of academic accreditation for degree courses. Now over 450 courses have been accredited in the UK, Hong Kong, Singapore, the West Indies and South Africa, with mutual accreditation agreements with

Australia, New Zealand and Ireland. In 1987 the Institution started a new system of accreditation of training schemes for students and graduate trainees, being the first institution to insist on training panel visits for all schemes. Now over 300 training schemes have been accredited in the UK, Hong Kong, Dubai and the Czech Republic.

The IEE therefore gave very careful consideration and debate, over a period of two years, as to how best to introduce a system for CPD. Should it be an accreditation system and, if so, on what basis? In the end, after looking at many other CPD schemes already operating in other institutions and professions, the Institution decided to introduce a voluntary policy of CPD from 1 January 1995 for a trial period of three years, with many members of the Institution's Council expressing the hope that it would thereafter become mandatory.

Three important aspects of the IEE approach are:

- it is open to all members;
- participation in the scheme is voluntary;
- the Institution will keep information on its database showing the amount of CPD each member has undertaken each year.

Mike Smythe, director of professional services, said at its launch: 'Our members are professionals, who are proud of their membership qualification. But it only shows an achievement at a particular point in time, namely when it was awarded. Showing proven CPD achievements thereafter will enhance our members' marketability.'

A points based system

Each participating IEE member who attains a minimum of 60 points of assessed CPD in any period of three consecutive years, on a rolling basis, will qualify for the award of a CPD certificate (or a CPD certificate of merit for 120 points, with their names published in 'IEE news').

In order to keep the scheme relatively simple to operate and control, only activities requiring attendance are included in the scheme in its present form. These fall into two types, which are:

(i) *Technical events*—aimed either at developing a member's specialised technical knowledge or at broadening this beyond the specialisation.

(ii) *Nontechnical events*—in this category come management techniques, interpersonal skills, employment law, health and safety law, finance, languages, etc.

At least twenty five per cent of a member's CPD activities must be in each of these two categories to qualify for a certificate.

The points allocation for all such events is:

- one point per hour of examined CPD;
- three-quarters of a point per hour of interactive CPD;
- half a point per hour of attendance only CPD;
- five points for presenting a one hour lecture or a refereed paper or article.

Various types of approved providers are:

- the IEE itself, including its centres, divisions, professional groups, distance learning committee, conferences, vacation schools and short courses;
- other professional institutions;
- higher educational institutes offering IEE accredited degrees;
- employers operating IEE accredited training schemes;
- others who seek and obtain approval from the IEE for their CPD provision. These could include training organisations, colleges of further and higher education, and employers not having IEE accredited training schemes.

Approved providers are required to operate within specified criteria and to appoint a senior person, known as the nominated individual and acceptable to the IEE, to act as the named link between the IEE and the provider.

Members are able to recognise IEE assessed CPD events by logos shown on the publicity material, or by a plain language statement such as 'this event qualifies for half a technical IEE CPD point'. A record of all assessed CPD provision is maintained in the IEE's Courses Information Unit in the library at Savoy Place.

Ken Smith, the IEE's CPD manager, says: 'This scheme will cost the Institution around £2 per member per year, plus our start up costs. And we think it is worth every penny!'

However, it is being increasingly realised that a points-based scheme such as this is only a step forward in CPD. The IEE is now working closely with the Institution of Mechanical Engineers to move towards a competence-based system.

Some future challenges

Professor Philip Secker, Deputy Secretary of the IEE, asks the question: 'Where do we go from here? Our scheme will operate initially for a period of three years and will then be reviewed. In the meantime, however, there is still a lot to do to develop the scheme further, both within and outside the IEE.'

Within the Institution he sees possibilities of extending the range of qualifying activities to include other forms of CPD, such as private study and on the job learning. Outside there is the work of The Engineering Council in developing the code of practice in CPD.

He would like to see the larger institutions helping and supporting the smaller ones to develop and introduce their own CPD schemes. To this end, the IEE is prepared to offer its scheme for free to any other institution wishing to adopt or adapt it; and even prepared, for a modest fee, to manage such a scheme on their behalf if so asked.

'Let me make three challenges,' says Professor Secker. 'First, we should find a way of achieving mutual recognition of CPD across the engineering profession. Secondly, let us start talking now about the creation of a mutually acceptable procedure for the assessment, measurement and recognition of each other's CPD. But my third challenge is broader still. After all, engineers will perhaps need CPD in law, in finance and in medicine, and will attend courses provided by these professions. So my final challenge is this: let us set up an inter-profession group to discuss ways in which we can recognise CPD activities from each other's professions.'

Let us therefore take up Philip Secker's point and look at the approach to CPD in the legal, accountancy, medical and insurance professions.

11.1.2 CPD and the Law Society

In view of the considerable debate in the engineering profession concerning mandatory CPD, it is particularly interesting to look across at colleagues whose status in society is often envied.

The equivalent for the solicitors to the engineering profession's Finniston Report in 1980 was the Royal Commission on Legal Services chaired by Sir Henry Benson in 1979. Initially, the Law Society introduced a requirement for all solicitors admitted as from August 1985 to undertake CPD for the first three years following admission to the profession. It was hoped that the activity would be habit forming, with the need to accrue a total of 48 continuing education points over the three year period, as well as to attend one half-day course in each year. In February 1990 it was decided that, with effect from September 1990, those solicitors admitted after July 1987 had to collect 16 CPD points for each year of their career and not just for the first three years.

In May 1990 the Law Society's council agreed that, subject to phasing arrangements to be agreed, the whole profession should be required to undertake the minimum training requirement of collecting 16 CPD points per year for the rest of their career. It was subsequently agreed that solicitors admitted after October 1982 would have to meet this CPD requirement each year, with effect from November 1994, and the rest of the profession would be included from November 1998.

In the meantime, there had been a tremendous growth in the number of organisations offering both formal courses and in-house CPD, as well as developments such the Legal Network Television. In November 1992, therefore, the points system was changed to an hours-based requirement of 48 hours over a three year cycle, except for the first three years after admission during each of which 16 hours of CPD must be undertaken.

The aim is to encourage the 60 000 solicitors to learn new skills and keep up to date with new areas of law, as well as to refresh all their skills such as researching. Among the requirements to be met are:

- at least twenty five per cent of this CPD activity must be met through attendance on courses offered by Law Society authorised course providers;
- no more than twenty five per cent of the CPD may use authorised audio/visual distance learning courses. This twenty-five per cent can also include writing law books, legal articles, legal research or a dissertation leading to a recognised qualification;
- there are restrictions on the amount of unsupported distance learning.

In addition, solicitors can add twenty five per cent to the actual time for attendance on workshop courses requiring delegate participation. Furthermore, solicitors who lecture on courses leading to qualifications and/or CPD courses can count twice the actual time on the first occasion on which they present the course.

Solicitors are also being required to keep a personal training record up to date. As from early 1993, the Law Society has been able to request sight of training records at any time and this has shown that the vast majority are complying with the requirements. The Law Society approach at present is that those not complying are talked to and persuaded of the importance of CPD. The ultimate sanction eventually might be for the annual practising certificate not to be renewed—in an extreme case where, for example, a solicitor had misrepresented the amount of CPD undertaken.

A 1993 discussion paper by the training committee of the Law Society suggested that at least twenty five per cent of the CPD requirement should be fulfilled by training in practical skills such as communication and other aspects of management development. This is currently a recommendation.

11.1.3 CPD in the Institute of Chartered Accountants in England & Wales

Next let us look at the approach to CPD in one of the large accountancy institutes.

The Institute of Chartered Accountants in England & Wales (ICAEW) considers that all members owe it to themselves, and their fellow professionals, to ensure that they are professionally up to date and that the reputation and value of their qualification is safeguarded.

The Institute expects all members to demonstrate a commitment to continuing professional education (CPE), and there are specific groups of members for whom this is compulsory:

- those applying for fellowship are required to confirm that they have complied with the guidelines on CPE on an annual basis;
- those supervising training in training offices and post qualification training offices authorised by the ICAEW;
- those members working in the reserved areas of audit, investment business and insolvency;
- those seeking entitlement to practise.

Because of the wide range of professional activities of its membership, the ICAEW relies on members and their firms to decide the relevance and usefulness of any CPE programme to their own circumstances. Normally CPE should be the assimilation of technical knowledge. It is recognised, however, that it may be appropriate to include time spent on the development of interpersonal or management skills.

A distinction can be made between unstructured and structured CPE. Structured CPE can be achieved through interaction with other individuals (not necessarily other members); for example, through attendance at technical meetings, seminars, lectures, courses (including pre-course/meeting preparation), as well as distance learning where the course is assessed and/or leads to a further qualification. Unstructured CPE is normally achieved through private reading and study.

To help in judging and assessing CPE achievement, a points system is recommended. A reasonable target is expressed as an average of 150 points year on year—with three points being recorded for each hour of structured CPE and one point for each hour of unstructured activity.

Although there is no minimum requirement for structured CPE, it is recommended that members for whom CPE is compulsory should aim to achieve at least forty per cent of their CPE (60 points) through structured activities (i.e. three days).

Recording and reporting CPE

Members for whom CPE is compulsory are required to maintain their own annual record of CPE undertaken and to be able, when required, to confirm to the Institute that they have complied with the CPE guidelines. If this has not been the case, it will be up to these individuals to explain the reasons behind their failure. In addition, members should be prepared to explain the relevance of their CPE to their personal professional development.

Members for whom CPE is not compulsory will still be expected to record their compliance with the CPE guidelines on an annual basis and, if necessary, be able to provide details of CPE undertaken and explain the relevance of that CPE to their professional development.

11.1.4 The right prescription for CPD

The College of Pharmacy Practice has some 800 members, all of whom also belong to the Royal Pharmaceutical Society of Great Britain. The latter has about 40 000 members and acts as the registration body for all pharmacists; registration requires a degree in pharmacy, one year's experience in an approved place of training with an approved tutor and the passing of the Society's exams.

In 1981 the College was formed with the support of the Society and became independent in 1985. Its aim is to raise standards even higher and has the Mission Statement:

> 'To promote professional and personal development, through education, examination, practice and research, benefiting patients and health care provision.'

Members of the College are formally required to undertake a minimum of 20 hours of continuing education per annum, as part of their CPD, as evidence of a commitment to the improvement of personal performance and standards of practice. (In contrast the Society recommends a minimum of 30 hours of CPD each year.) College members must complete a simple form each year listing:

- accredited postgraduate courses or meetings;
- accredited distance learning courses;
- alternative methods used to satisfy the 20 hour plus requirement.

As a result, some members have withdrawn their membership, but others have joined precisely because of this mandatory policy.

The membership examination is designed to measure high standards of pharmacy practice using assessment methods suitable for mature professionals and reflecting the candidate's own practice.

Accreditation of courses

In order to encourage higher standards and greater professionalism in the organisation of courses, the College has, since January 1991, operated a scheme to accredit the providers of courses considered suitable.

In addition to this continuing education requirement, the College encourages its members to develop their CPD activities outside pharmacy, for example in law, computing and management.

Professional development programme

There is a regular programme of study days, highlighting new developments such as professional audit, partly financed by the participants, who pay £30 per day, with the rest sponsored by industry. There are also many courses run by National Funded Centres financed by the National Health Service.

Credit for learning through reading

One particularly innovative approach which has been launched by the College is the provision of multiple choice questions relating to specified articles appearing in 'The hospital pharmacist', published bi-monthly by the 'Pharmaceutical journal', which is available to all hospital pharmacists. 'After studying the relevant article,' explains Rosemary Mitchell, the College's administrator, 'pharmacists will be able to answer the questions on the paper, returning their questionnaires to the College for marking within three weeks. They will then receive a detailed assessment of performance, notes on the answers (which will be published in the next issue, together with these explanatory notes) and a certificate of completion which will count towards their continuing education requirement. Likely subjects will be topics relevant to the science, technology or practice of pharmacy, as well as relevant management topics.'

Each assessment will carry an indication of the length of time required to read and study the article, look up any references and complete the assessment. This will probably range from one half to two hours, depending on the length and complexity of the article. Over one year, it would be reasonable to expect that about six hours of continuing education would be achieved in this way.

11.1.5 A healthy approach to CPD

Next we will look at the approach to CPD in another part of the health service—that of nurses, midwives and health visitors.

The United Kingdom Central Council (UKCC) is the statutory regulatory body for 639 000 nurses, midwives and health visitors in the United Kingdom. Midwives have had a statutory requirement for continuing professional development for many years in the form of refresher courses and return to practice programmes. Since April 1995 these requirements have been expanded to include all registered nurses, midwives and health visitors and apply to everyone re-registering, as is required every three years.

The new scheme, known as PREP (which arose from the post-registration education and practice project which formulated the requirements), affects every nurse, midwife or health visitor who wishes to maintain registration with the UKCC. Registration is a requirement for professional practice in the UK.

The scheme has three parts, some of which are statutory and some of which are guidelines of good practice. The three parts relate to:

- a period of support under the guidance of a preceptor during the first four months of professional practice—good practice guidelines;
- requirements for maintaining registration—statutory requirements;
- requirements for specialist and advanced practice—statutory requirements.

Let us look at each of these three areas.

A period of support under the guidance of a preceptor

The guidance of a preceptor is recommended for four months to assist newly registered nurses or midwives. This has been Council policy since 1993 and there is evidence of an increasing number of excellent schemes being put into place to offer newly-qualified staff such support. There are of course strong parallels here with the use of mentors in engineering training schemes.

Requirements for maintaining registration

All individuals who are working in any capacity with a nursing, midwifery or health visiting qualification are required to maintain registration, including those who are in teaching, management or policy making, as well as those involved in direct hands on patient or client care.

In essence, the requirements are designed to help individuals to maintain and develop their professional knowledge and competence. The system is based unequivocally on the concept of individual practitioner accountability. The requirements are made known to each individual at the point of re-registering, on a three-yearly basis, and are that each practitioner must:

- complete a minimum of five days of study activity every three years;
- complete a notification of practice form to inform the Council of the qualifications which they are using for practice;
- use a personal professional profiling system to identify learning and professional career needs, as well as the benefits which have accrued to patients and clients as a result of the additional learning;
- complete a return to practice programme where individuals have been out of professional practice for five years or more.

The Council will not be offering any formal system of approval for study activity. Individuals will choose activities which are relevant to their professional registration and their role. Individuals are given the following categories from which to choose:

- patient, client and colleague support;
- care enhancement;
- practice development;
- reducing risk;
- education development.

Individuals are encouraged to review their personal professional competence and identify their strengths, weaknesses and the areas in which they wish to develop. They are then asked to identify what they wish to achieve and select appropriate learning activities to meet their needs. This could include, for example:

- a literature search;
- a seminar;
- attendance on a course;
- a visit to another unit or centre of excellence or specialist expertise which is relevant to their practice.

Individuals are then invited to consider what they have actually learned as a result of implementing their action plan. This would include consideration of the strengths and weakness of the activity, event or visit, and whether it enabled them to meet their objectives. The value of what has been learned for both patients, clients and colleagues, and how the knowledge has been shared, will also be recorded.

In due course (after 1 April 2001, by which time everyone will have completed their first three years on the scheme, including those whose re-registration only comes up in March 1998) a formal audit system will be operated by the Council to ascertain how its requirements have been met.

Requirements for specialist and advanced practice

The Council recognises that pre-registration preparation ensures competent practitioners at the point of registration. It further recognises that individuals will add to their knowledge and skill over the years. However, specific preparation is also required for specialist practice. Programmes leading to a specialist qualification will be of an academic year in length and of degree level outcome. Individuals who meet the Council's requirements in respect of the learning outcomes for the particular programme will be entitled to hold that qualification against their name on the Council's register. Further work is currently being undertaken within the area of advanced practice—the intention being to define this more exactly and to identify appropriate standards.

Funding

Individuals will be expected to fund their own continuing professional development activity where employer support is not available. However, the four government health departments have been very positive in supporting this initiative and recognise that, although the responsibility for maintaining registration lies with the individual practitioner, employers also have a responsibility and are encouraged to ensure that their staff do meet at least the minimum requirements.

Reaction by the professions

The professions have recognised and welcomed the PREP proposals. Understandably, there are some concerns about individual aspects of the recommendations, but there is a strong and growing movement of commitment towards lifelong learning and continuing professional development.

11.1.6 CPD for personnel and training professionals

The Institute of Personnel and Development (IPD) launched its new CPD policy in 1995. As the sole UK professional institute in the field of people management, the IPD is looked to by employers as a primary source of information and advice about CPD and considerable thought has gone into the organisation's CPD policy. For the previous three years, applicants wishing to upgrade to corporate member or fellow status had to demonstrate that they had been keeping themselves up to date through CPD.

The CPD requirements of the IPD are:

- the most important aspect of CPD is the outcome, not the precise amount of input. In order to keep up to date, the minimum level of CPD activity is recommended as 35 hours per year or the equivalent in learning hours/outcomes;
- the IPD will expect records to reflect a balanced mix of activities. These should include: professional work-based activities, courses/ seminars/conferences and self directed/informal learning;
- through reflection and the process of completing a record (one option being a computer-based program developed by IPD) further development needs should be identified. Each individual should have their own development plan identifying further CPD needs.

'Members will also get a lot of help from their branches, which have all appointed CPD advisers for whom we held our first CPD conference in October 1995,' says Christine Williams, membership development manager.

11.1.7 Managing to improve CPD

The Institute of Management (IM) launched its CPD policy in September 1995, encouraging individuals to take charge of their own development. IM represents over 70 000 individual members, making it the largest broadly based management institute in the UK. It also embraces 700 corporate members, representing around three million employees. The approach endorsed by the IM Council emphasises that CPD must be centred around the individual. Each must accept responsibility for assessing and meeting their own needs, using the help available from their professional organisations, educational institutions and employers.

A CPD pack has been made available. This allows managers to develop their own competencies and support their career development. Comprehensive support materials include self assessment tests, enabling managers to evaluate their development needs and check their progress. The pack also includes a career long personal development record, as well as up to date information on IM publications and courses and other centrally organised activities.

Roger Young, the IM director general, comments: 'Skills constantly need to be updated and improved. Individuals must be prepared to invest in their own development. CPD is all about the practice of maintaining one's professional development throughout a career and considering that the current state of skill or knowledge is never sufficient. All new members of the Institute commit themselves to this policy when they join.'

This view was supported by Michael Heseltine, the then Deputy Prime Minister and First Secretary of State who, commenting on the new policy, said: 'In a world where the rate of change is ever increasing, training throughout working life is now essential both for individuals and for the UK's competitiveness.'

11.1.8 Insuring for the future

One other profession that we can look at in the UK is the insurance industry.

The Chartered Insurance Institute (CII) has introduced a scheme, as from 1 January 1995, specifying that its 15 000 chartered members must complete a minimum amount of CPD each year in order to retain chartered status. This is in response to concerns that skill levels in industry and commerce in the UK are not, in general, what they should be, and a need to consider the international implications of this lack.

'At one time,' says Clive Sanderson, the CII's CPD project manager, 'a qualified insurance man or woman was almost set up for life, once professional qualifications had been achieved, without necessarily having the need for ongoing development. How this has changed! It is now not sufficient to be qualified—it is also necessary to keep up to date with rapid change as a direct consequence of competition. New skills need to be learnt and developed.'

The new CPD scheme applies to all UK and Ireland based chartered insurers and chartered insurance practitioners. Activity is specified as divided between structured and unstructured CPD, with points awarded as follows:

- structured (three points per hour)—courses, conferences, research, lecture preparation, technical authorship, assessed distance learning, exam setting, etc.;
- structured (two points per hour)—professional or trade body activities, such as attending institute meetings or acting as a mentor, examination marking and moderation, and developing specialist technical knowledge at work, such as by the development of new products, systems or corporate strategy;
- unstructured (one point per hour)—unassessed distance learning, videos, television programmes, audio tapes, CBT, reading technical digests, bulletins, manuals, journals, etc.

It is emphasised that this activity, whether structured or unstructured, must be relevant to the individual's own development needs—and only he or she can judge that. Consequently, the scheme is self certifying, but requires that:

- a total of 180 points must be achieved over a three year period;
- for 1995, up to seventy five per cent may be unstructured;
- for 1996, up to fifty per cent may be unstructured;
- for 1997, and thereafter, the maximum unstructured CPD will be set at twenty five per cent;
- between 20 and 30 hours of CPD are undertaken each year as a minimum.

The CII will regularly call in the CPD records of those with chartered titles on an annual basis, as well as the records for all new holders within a given calendar year, as a check that the CPD requirement is being carried out.

There is also a system of accreditation of commercial suppliers of structured training with random checks of individual courses and materials.

The objectives

The CPD scheme is open to the remaining members of the Institute (67 000 in total) on a voluntary basis if they so wish. Other key objectives are to:

- ensure that professionals keep knowledge up to date, particularly in their own specialist areas;
- encourage public confidence in the holders of descriptive titles;
- consolidate learning from experience gained at work;
- be aware of influences in other branches of the industry;

- develop new skills, either to improve current effectiveness or for a future, more responsible, role.

A major financial investment

One indication of the seriousness with which the CII is taking CPD is the establishment of a CPD unit in Woodford, with an initial budget of £600 000, and five staff. This is designed to support CPD initiatives in liaison with local institutes, the Society of Fellows (open to Fellows of the CII), the Society of Financial Advisers and the Society of Technicians in Insurance.

The main aim of the CPD unit is to ensure that adequate CPD provision is available on a nationwide basis, as well as to ensure effective co-ordination and the maintenance of standards of CPD programme delivery. As part of these objectives, the unit will provide funding in a number of project areas, including guaranteeing programmes against financial loss, commissioning certain CPD material and sponsoring events such as videoconferencing important lectures which could not be held in other parts of the country, and sponsoring research projects.

Clive Sanderson summarises this CPD scheme by saying: 'As can be seen, the CII is very committed to this project. It is vital that the scheme succeeds, and this can only happen with the full support of our membership. Training and development are vital to the future of our industry and to our competitive position as a country. An emphasis on self development (rather than putting all of the onus on employers) is essential and perhaps long overdue. CPD now provides a golden opportunity for such development!'

CPD is indeed a way of insuring your own future. Let us return to see how some more UK-based engineering institutions are developing their CPD schemes.

11.1.9 The Institution of Mining and Metallurgy

At its December 1994 Council meeting the Institution of Mining and Metallurgy approved a definition of CPD and a code of practice. This definition is: 'CPD is the on-going enhancement of an individual's professional qualities and expertise to establish the highest standards, technically, managerially and commercially. It should develop:

- informed judgement;
- a readiness to keep abreast of the latest technology and business practices;

- an attitude of self reliance;

and thus contribute to:

- individual and corporate capability;
- anticipating and preparing for future change.'

The code of practice is as follows:

1 The Institution of Mining and Metallurgy expects that individuals will undertake CPD to maintain their professional standards.
2 The individual's professional development should seek to meet the objectives set out in the Institution's definition of CPD.
3 Individuals seeking election or transfer to associate, member or fellow of the IMM will be expected to provide information of CPD activities they have undertaken.
4 The Institution will provide guidance on acceptable forms of CPD and on desirable means of recording this.
5 From time to time the Institution will review its policy on CPD and advise members of changes, as appropriate.

11.1.10 Mining the benefits of CPD

The Council of the Institution of Mining Engineers has adopted a policy for CPD and a code of practice for all members. The policy requires that:

- each member of the Institution, other than those who have retired and no longer practise their profession, must maintain their professional competence by identifying their needs, planning suitable action and reviewing their achievements. Suitable action can include the presentation of original papers, attendance at courses, lectures, training sessions, innovative experience or secondment and self-regulated learning;
- each member, other than those who are retired, is obliged to spend at least 30 hours each year, averaged over the previous three year period, on professional development. The member is expected to include different forms of CPD activity;
- the development activities must be recorded in such a way that each component can be identified and assessed. The Engineering Council's 'Career manager' can be used for this purpose.

The most important aspect which will affect applicants and transferees is that, as from 1 September 1995, evidence of meeting the Institution's CPD requirements has been looked for at the professional interview for election or transfer to the grade of fellow, member or associate member of the Institution.

11.1.11 Mechanicals launch CPD

In January 1995, the Council of the Institution of Mechanical Engineers (IMechE) formally adopted a CPD policy. This states that all members who wish to remain recognised as practising professionals should feel obliged to undertake a structured programme of CPD; unless individuals are able to demonstrate this to their present and prospective employers, then they will become progressively less attractive in an increasingly volatile labour market. The stipulated time to be devoted to CPD is an average of 50 hours annually.

The IMechE stresses the complementary values of updating knowledge, deepening knowledge in the individual's specialist area and broadening knowledge both across a wider engineering spectrum and in general life skills such as management, leadership and finance. A wide variety of subjects and learning vehicles is recognised, with activities ranging from formal courses and seminars to the less formal, including private reading and on the job experiences such as putting together a budget for the first time—provided that the activity can be verified by a competent third party or mentor. The criterion for what represents valid CPD is whether the activity or subject is relevant to the individual's and/or the employer's needs.

As from 1994, the IMechE has issued a CPD guidance booklet free of charge on request, and provides a personal professional development record for a modest cost. With the formal launch of the scheme, these will be reinforced by the provision of a more comprehensive advice service to members and their employers. One of the IMechE's aims is to produce further guidance, such as statements of the key competencies required to perform well in each of a range of employment sectors; members will then be able to review at a glance the subjects and skills which need attention. A pilot based on competencies was launched jointly with the IEE in January 1997; some participants used a paper-based set of documentation and others used the PDP on the Web software described on page 135.

A programme of regional launch activities has also been planned, for the benefit of members and employers; the IMechE recognises the need for CPD to be supported on a local basis and encourages members to take advantage of local activities, including those run by other institutions.

CPD enthusiasm too from IMechE Young Members

The Institution of Mechanical Engineers' Young Members have for some time been formulating policies and facilitating CPD aimed specifically at the Institution's young membership.

In 1994, its Younger Members' national assembly was devoted to CPD and drafted policies, as well as suggesting initiatives which they have been pursuing since. A working party was established to concentrate on raising awareness of CPD and the first results were seen in 1995. The aspects concentrated on have been:

- raising awareness and generating enthusiasm within the Institution's regional Young Members' organising committees. Information packs advising on suitable activities have been produced and chairmen provided with presentation material to help them put forward the key ideas;
- encouraging committees to advertise their existing activities, where appropriate, as being suitable for CPD. Certificates of attendance are being issued so that, at a future date, such as when applying for corporate membership, members can prove what they have been doing;
- producing an information leaflet to raise awareness within the young membership. This was distributed to all Young Members and outlines what CPD is, why it is important and where they can pursue the necessary activities.

11.1.12 Joint BCS-IDPM initiative for the information technology industry

The two leading bodies of the IT/IS industry, the British Computer Society (BCS) and the Institute for Data Processing Management (IDPM) launched a national CPD programme for IT/IS professionals in 1995.

The programme enables BCS and IDPM members to have their professional updating activities recorded, assessed and validated, and its launch followed the succcessful completion of a joint scheme by the two professional bodies, which operated on a regional basis.

Those participating are able to record evidence of the maintenance and broadening of their professional skills, which may include any of the following activities:

- branch and special interest group meetings of either the BCS or IDPM;

- conferences/workshops/seminars;
- training courses;
- use of open and distance learning materials.

Jean Irvine, group IT director for The Post Office, and chair of the BCS Effective Leadership in Information Technology group, says: 'Rapidly changing technology creates significant challenges in keeping skills and professional standards up to date. The new BCS/IDPM programme for CPD provides us with a framework which will help to meet these challenges. It is now up to all of us in senior management positions in the industry to use it to the full.'

11.1.13 A civil approach to CPD

The Institution of Civil Engineers (ICE) has an obligatory CPD policy for its qualified members, and has been developing a more comprehensive recording and planning document to supplement the 'Record of CPD' (ICE108), which is still obtainable free from the ICE's training department.

In an obligatory system the ICE does not see the need to approve suitable activities. However, information is available on the ICE bulletin board under training courses and meetings which may be suitable for use as CPD. Access may be obtained via a modem or PSS link, and it is available 24 hours a day free of charge. A guide to the ICE's computer based information systems is available on request from the ICE's library.

For those wishing to progress either to corporate (MICE) or associate (AMICE) membership, 30 days and six days of CPD respectively, known as continuing education and training (CET), are required before taking the appropriate professional review. Guidance on the types of activities that may contribute towards this requirement is given in the relevant routes to membership documents. Emphasis is placed on the need to cover a balanced programme including technical, managerial/professional and safety issues. Demonstration of involvement in Institution affairs is also a requirement. The ICE organises approved training schemes with employers, and as part of this it is the responsibility of candidates to satisfy their supervising civil engineers, or employers or sponsors if not on an approved scheme, and in time their reviewers, that their CET has been useful and valid, and that they have benefited from the activities.

11.2 CPD and the future

Having looked at the approach to CPD in a wide range of both engineering and nonengineering institutions, we should ask where the professional institutions will need to take CPD in the years ahead? I offer the following suggestions:

- there will be a need to develop a common framework and language. This will enable, for example, engineers to gain accredited CPD from the medical profession and vice versa. We are all becoming much more multiskilled—and if we are not, then we jolly well ought to be doing so! All professions, with a common framework, could benefit equally from management courses, as well as benefiting from CPD internationally; a career move from one country to another would then enhance everyone's CPD, indeed as it should, rather than in any way restricting individuals to obtaining CPD credits in their home countries;
- much greater credit needs to be given to on the job learning. This is, after all, where most of us obtain the bulk of our learning;
- CPD will increasingly need to be based on maintaining, raising and broadening competencies. Competence accreditation, quite possibly by mentors, will become the key to validating CPD;
- software, I predict, will play an increasingly important role in both enhancing and measuring CPD, particularly in terms of measuring the audit trail of competencies of institution members. Chapter 8 explained more about these possibilities, but already the IEE, for example, is making considerable use of the Internet to give its members access to its publications, etc. Use of PDP software and groupware, such as Lotus Notes, will very soon become enormously important in CPD, particularly through the institutions;
- recognition will need to be given more and more to the fact that increasing numbers of individuals will work independently, marketing their expertise to a wide range of customers, and in some cases giving it for free to their institutions—just as I do! Institutions can greatly assist both by enabling these individuals to stay at the forefront of expertise in their specialisms—which is what gives them their market value, after all—and providing a CPD framework to enable them to demonstrate their experience and expertise to potential clients.

11.3 Summary

Professional institutions started as voluntary bodies, but increasingly both individuals and the public will see them as indispensable organisations for ensuring that the services provided by professionals are both of high quality and safe. Essentially, professional institutions operate by peer review—with groups of members with expertise in a particular area assessing the competence of other individuals—and it is this process which has to be at the heart of any quality system.

So, if you as an individual, employer or training provider are not already working closely with your relevant professional institutions, make the commitment to do so a very high priority in your personal development plan! It is only with such commitment from all concerned, after all, that institutions can be really effective—so in the end everyone gains.

Chapter 12
The international dimension

'By the year 2000 fifty per cent of the world's GDP will be produced outside OECD countries. We have no option but to raise our level of competence. The best solution is to have all round individuals rather than specialists'—François Cornelis, chief executive officer and vice chairman, Petrofina (keynote speech at First World Conference on Lifelong Learning, Rome, December 1994)

This book started with an international flavour—and that is the way I would also like to end it.

Since starting to write I have had the good fortune to travel in a professional context to a number of different countries, including Belarus, China, Dubai, Holland, Hong Kong, Italy, Finland, France, Brazil and Turkey. Using my connection to the Internet I have been able to communicate regularly with still more countries.

As the quote at the beginning of this chapter indicates, we all need to have an international perspective, even if we do not necessarily have the opportunity of regular international travel.

The IEE itself, for example, has 30 000 of its 140 000 members based outside the UK, with many other institutions also having a significant and important international membership.

12.1 The International Association for Continuing Engineering Education

One excellent way in which you can extend your professional network internationally is by joining the International Association for Continuing Engineering Education (IACEE).

This organisation developed out of the continuing engineering education (CEE) working groups of UNESCO. In 1989 it was decided to launch an independent organisation, IACEE, to take over the activities of these CEE working groups, including the organisation of the world conferences on CEE every three years. IACEE was launched at the 4th World Conference on Continuing Engineering Education in Beijing in 1989 and the IEE became one of the founding members, together with the Institution of Mechanical Engineers and the Institution of Civil Engineers. I was there in Beijing and found myself elected as one of the 16 council members, and then within that as one of the five executive committee members.

IACEE now has some 600 members from over 70 countries. There are four categories of membership:

(i)   professional organisations—these include the American Society for Engineering Education, the Japan Society for Engineering Education, the European equivalent SEFI and all the equivalents in other continents;
(ii) industrial organisations and companies;
(iii) academic institutions and other co-ordinators and offerers of CEE;
(iv) individuals.

If you are interested in joining, then contact:

Markku Markkula
Secretary General
IACEE
c/o Helsinki University of Technology
Lifelong Learning Institute
FIN-02150
FINLAND

Tel: +358 0 451 4028
Fax: +358 0 451 4060
e-mail: IACEE@hut.fi

Further details about IACEE, including the latest activities of the professional development working group, can be obtained through the world wide web on the IACEE home page at the address:

http://www.dipoli.hut.fi/org/IACEE/

IACEE provides many different forums for developing one's CPD such as:

- a high quality regular newsletter;
- publications on such topics as 'Lifelong learning of engineers in industry—report on lifelong learning practices in Europe, USA and Japan' and 'The role of education and training in the internationalization process of Japanese companies';

- the world conferences on continuing engineering education;
- providing pump-priming funds for research into CPD.

One of my visions through IACEE is to foster the concept of a continuing professional development programme (CPDP). CPD is increasingly an international activity—so surely we need an international framework to ensure that it develops as fast as possible!

At the fifth world conference in 1992, an international CPD working group, chaired by myself in my capacity as a council and executive member of IACEE, developed the concept of the CPDP. This will be a seamless framework covering both technicians and engineers, without time constraints.

An important part of this is a proposed international modular masters degree framework. At present there are a large number of self contained masters degree schemes. Examples in the UK are the Open University, Jupiter and TEAM. Across Europe there are the EuroPro and satellite based EuroPace 2000 schemes. In North America the National Technological University (NTU) satellite scheme has been highly successful. In the Commonwealth of Independent States IACEE has been introducing NTU and EuroPace modules through its East-West initiative.

Why not, therefore allow a student enrolling on one scheme to continue on, or take modules from, another scheme? The benefits would be:

- *to the providers*—more students, both nationally and internationally;
- *to the students*—the opportunity to extend the modules available to them from other schemes and other countries;
- *to the employers*—broader developed engineers in an international context.

If you wish to assist in the development of this scheme then please contact me through the IEE. Those wishing to participate in the professional development working group can do so through the IACEE home page given earlier.

12.2 SEFI

The Société Européenne pour la Formation des Ingénieurs (SEFI) is the French title for what in English is called the European Society for Engineering Education. It has a working group on continuing engineering education, the secretary of which is Anders Hagström, now based at Cambridge. At the time of writing the chairman is Dr. Christopher Padfield who heads the Cambridge Programme for Industry at Cambridge University, with the vice chairmen being Michel Futin at the École Centrale de Lyon in France and Alexander van den Eijnde of the Stan Ackermans Institute in Eindhoven, Holland.

Do give serious consideration to joining SEFI. The contact is:

Madame Françoise Côme
Secretary General
SEFI
60 rue de la Concorde
1050 Brussels
BELGIUM

Tel: +32 2 540 9770
Fax: +32 2 540 9715
e-mail: sefi.come@infoboard.be
world wide web: http://www.ntb.ch/SEFI/

12.3 FEANI

The Fédération Européenne d'Associations Nationales d'Ingenieurs (FEANI) brings together national engineering associations from 22 countries as national members. It was founded in 1951, with head-quarters located in Brussels at:

FEANI
21 rue du Beau Site
B–1050 Brussels
BELGIUM

Tel: +32 2 639 0390
Fax: +32 2 639 0399

FEANI's aims include:

- securing the recognition of European engineering titles and protecting those titles, in order to facilitate the freedom of engineers to move and practise within and outside Europe;
- safeguarding and promoting the professional interests of engineers;
- fostering high standards of formation and professional practice, and encouraging the regular review of these;
- promoting cultural and professional links within the engineering profession, especially in Europe.

For these purposes FEANI maintains a register to which individuals may be admitted, provided that they meet the specified minimum requirements. The minimum education standard is defined as:

$B + 3U$

where B represents a high level of secondary education validated by one or more official certificates awarded at about the age of 18 years, and U represents a year (full time or equivalent) of approved engineering education either given by a university or other recognised body at the university level included in the 'List of schools and courses' accredited by FEANI.

The minimum standards for registration on the basis of formation (the total of education and initial training), allowing someone to put the title Eur Ing, for European engineer, before their name is:

$B + 3U + 2 (U \text{ and/or } T \text{ and/or } E) + 2E$

where T represents a year (full time or equivalent) of training through a programme, the aim of which is to increase knowledge through work within technical fields, for instance on a construction site, in a factory, laboratory, office or other working environment, defined, supervised and approved by a university or by a body accepted by FEANI as part of engineering formation, and E represents a year (full time or equivalent) of relevant engineering experience assessed and approved by FEANI.

These details are given simply to illustrate that gaining common agreement across Europe on who should be recognised as a professional engineer has not been easy. Europe has a number of different traditions in both the education and training of engineers, and reconciling these took many years of discussion. In simple terms, however, if you are a chartered engineer in the UK, then you should have no difficulty in obtaining the Eur Ing title, and this is likely to become increasingly important for those developing their careers in continental Europe.

UK engineers applying to FEANI for the title need to apply through The British National Committee for FEANI, which is based at The Engineering Council in London.

It is worth noting that FEANI produces a CPD newsletter, published twice a year by the Finnish Association of Graduate Engineers, TEK, in Helsinki. For details of this contact:

Ms Sirkka Poyry
Director, Technology and Society
The Finnish Association of Graduate Engineers TEK
Ratavartijankatu 2
FIN-00520 Helsinki
FINLAND

Tel: +358 0 159 0314
Fax: +358 0 159 0306
e-mail: Sirkka.Poyry@tek.fi

12.3.1 The FEANI code of conduct

The FEANI code of conduct is additional to, and does not take the place of, any code of ethics to which the registrant might be subject in his or her own country.

All persons listed in the FEANI register are obliged to be conscious of the importance of science and technology for mankind, and of their own social responsibilities, when engaged in their professional activities. They exercise their profession in accordance with the normal rules of good conduct of European societies, respecting particularly the professional rights and the dignity of all those with whom they work.

They thereby undertake to comply with and maintain the following code of ethics:

1 *Personal ethics*
- the engineer shall maintain his/her competence at the highest level, with a view to providing excellence of service in accordance with what is regarded as a good practice in his/her profession and having regard to the laws of the country in which he/she is working;
- his/her professional integrity and intellectual honesty shall be the guarantees of his/her impartiality of analysis, judgment and consequent decision;
- he/she shall consider himself/herself bound in conscience by any business confidentiality agreement into which he/she has freely entered;
- he/she shall not accept any payment except those agreed with his/her relevant employer;
- he/she shall display his/her commitment to the engineering profession by taking part in the activities of its associations, notably those which promote the profession and contribute to the continuing training of their members;
- he/she shall use only those titles to which he/she has a right.

2 *Professional ethics*
- the engineer shall accept assignments only within the area of his/her competence; beyond this limit, he/she shall seek the collaboration of appropriate experts;
- he/she is responsible for organising and executing his/her assignments;
- he/she must obtain a clear definition of the services required of him/her;
- executing his/her assignments, he/she shall take all necessary steps to overcome any difficulties encountered, while ensuring the safety of persons and property;
- he/she shall take remuneration corresponding to the service rendered and the responsibilities assumed;
- he/she shall try to ensure that the remuneration of each be consonant with the service rendered and the responsibilities assumed;
- he/she strives for a high level of technical achievement, which will also contribute to and promote a healthy and agreeable environment for his/her colleagues.

3 *Social responsibility*—the engineer shall:
- respect the personal rights of his/her superiors, colleagues and subordinates by taking due account of their requirements and aspirations, provided that they conform to the laws and ethics of their professions;
- be conscious of nature, environment, safety and health and work to the benefit and welfare of mankind;
- provide the general public with clear information, only in his/her field of competence, to enable a proper understanding of technical matters of public interest;
- treat with the utmost respect the traditional and cultural values of the countries in which he/she exercises his/her profession.

12.4 The European Union Directive

On 17 April 1991 the European Commission Directive 89/48/EEC came into force. This is designed to ease the movement of professionals across Europe and is concerned with the mutual recognition of professional qualifications.

Information on this directive is available in the UK in a Department of Trade and Industry booklet entitled 'Europe open for professions'.

12.5 ELLI

The European Lifelong Learning Initiative (ELLI) was formed in 1993 and now has a network of about a thousand organisations.

Its mission is to provide an employment-driven focus on lifelong learning. It provides definition, direction and leadership in understanding and supporting European policies, structures and attitudes, encouraging and enabling the development of lifelong learning across Europe.

ELLI, for example, co-ordinated the First Global Conference on Lifelong Learning in Rome towards the end of 1994. The European Commission, the Council of Europe, the Club of Rome, UNESCO and the American Council on Education, were among some 500 delegates participating from around the world. As a result the World Intiative on Lifelong Learning (WILL) was formed, as well as ELLI networks specific to many European countries. WILL has now developed regional membership organisations equivalent to ELLI in every continent of the world.

It is always useful, in my view, to belong to organisations like IACEE, SEFI, ELLI and WILL, all of which work together closely. They extend your network of contacts, keep you up to date with the latest ideas and activities, and provide valuable newsletters, seminars and conferences. For example, I had the privilege in 1994 of attending an OECD seminar on lifelong learning in Japan as one of the two representatives of ELLI.

For further details of ELLI contact:

Keith Davies
President of ELLI
rue de la Concorde 60
B-1050 Brussels
BELGIUM

Tel: +32 2 540 97 52
Fax: +32 2 514 11 72
e-mail: 100344.2376@compuserve.com

It is also worth noting that the subject of lifelong learning is sufficiently important that the European Commission decreed that 1996 be the European Year of Lifelong Learning.

Two reports that the European Commission has published and which are worth reading are 'Skills shortages in Europe'[14] and 'Quality and relevance—the challenge to European education'[15] (both are free from fax number + 32 2 295 4361). Both reports were the result of many meetings between many of the most eminent industrialists across Europe, and each report has been widely praised for its wealth of case studies and practical facts and advice for action. The working group which produced the first of these reports was chaired by Sir Robert Telford, life president of the Marconi Company and a very distinguished British engineer.

Seeing CPD in an international context is very important. I will now reproduce four articles from The Engineering Council's 'CPD Link' newsletter which may give you food for thought in whatever capacity you are reading this book.

12.6 Removing the wall

Dr. Frank Huband, executive director of the American Society for Engineering Education (ASEE), proposes a national programme of CPD for the United States. Writing in the editorial of the society's magazine 'ASEE prism', Dr. Huband compares the engineer's career to a marathon run.

'Marathon runners know the feeling well. At about 20 miles, they hit the wall, an endurance barrier which they must overcome if they are to compete successfully. To break through the wall, runners must add nourishment along the way and call upon training for an accumulated reserve of will.

'An engineering career is a marathon, but the burden is greater. The average engineer hits the wall about 20 years after college and tends to level off in earning power. To negotiate this wall, engineers must constantly add nourishment in the form of continuing professional development. Employers, in turn, must realise that although it may seem less expensive and easier to gain current technical expertise by replacing mid-career engineers with younger ones, it is a false saving. By not providing their engineers with the opportunity to stay current, companies are foregoing the chance to develop the knowledge and depth of experience which only mature engineers with continuing education can provide.'

Dr. Huband argues that undergraduates need to be made aware that they may encounter a barrier as they enter middle age and that they should stay technically current, and states the need in the US to work towards an effective national programme for CPD.

'Universities and corporations could share the costs of a national program. Working together, using all available training resources, could enable even smaller employers to provide full, technologically advanced updating to employees. By interfacing with industry, engineering faculties would be challenged by mature, experienced engineers and could use what they learn in undergraduate classrooms and laboratories. Industry would gain constant access to cutting-edge technical research in the academic arena.'

12.6.1 Mandatory CPD in the USA

Around about 1980, the Iowa Engineering Registration Board broke new ground by mandating that engineers registered in that state demonstrated continuing professional proficiency. The Iowa law was crafted to require that registered engineers participated in continuing professional development, and certified to the Board at each renewal that the required amount had been carried out. Spot checks, rather than the checking of each individual record, are used by the State Board for enforcement.

The State of Alabama has passed a similar requirement for mandatory CPD and this law started being implemented in 1993. An additional ten states have pending legislation or plans to move in this direction.

The driving force for insisting on CPD comes from public concern that licensed professionals of all types stay up to date. Most of engineering's sister professions—from law and medicine to nursing and accounting—have laws for mandatory proficiency updating in almost every state of the USA. If one looks at a table of each of the professions licensed in all 50 states, the only column with a large number of blanks is engineering. This leads to concern that public pressure will soon demand of engineering what it has long demanded of the other professions.

In anticipation of the need for registration boards in many states to move in the direction of mandatory CPD for engineers, the National Society of Professional Engineers has interacted with the National Council of Examiners for Engineering and Surveying to develop a model law which would be available for state registration boards to consider. The hope is that, when pressed to move in the direction of mandatory CPD, the states would at least be adopting a law identical to that in other states. Such uniformity would make it easier for engineers registered in multiple states to meet the requirements.

As currently drafted, the model law would require each registrant to obtain 24 professional development hours (PDH) each calendar year. A PDH is defined as a contact hour of instruction or presentation, and could be earned by participation in college courses, continuing education courses, video short courses, industry seminars or workshops, conventions, etc. A PDH could also be earned through authoring publications, participation in professional organisations, peer review process and approved active practice.

12.6.2 CPD down under

The Institution of Engineers, Australia, strongly encourages professional engineers to involve themselves in regular professional development programmes. The Institution maintains a National Professional Engineers Register, which includes only those members who have been prepared to certify each year that they have undertaken sufficient CPD to meet the established Institution criteria. The current policy for continuing education reflects an expectation that professional engineers should complete at least 150 hours of professional development in any three year period. This may include formal, structured courses undertaken through face to face or distance learning methods, publication, presentation of lectures and structured private study.

Members able to comply with this requirement are issued with a practice certificate. Statutory registration of professional engineers applies only in the State of Queensland, so the National Professional Engineers Register fills a national need. The decision has recently been taken to strengthen the NPER by making involvement in relevant CPD mandatory and auditable for all members requiring certification of their competence in an area of engineering expertise.

The provision of CPD to the 57 000 members of The Institution of Engineers, Australia, in Australia itself, and a further 6000 members in many other countries within and beyond the region, representing every major engineering discipline, has some very special challenges. Many of the members are located in remote areas and this, along with the almost universal demand for flexible self-paced learning opportunities, effectively mandates distance education as a primary element in CPD activities.

The Institution thus has a firm commitment to supporting, encouraging and recognising CPD, and has established a subsidiary company, Engineering Education Australia Pty Ltd, to source funds, and develop, market, manage and evaluate distance education activities. The primary aims of EEA are to support individual engineers seeking to maintain and enhance their knowledge and skills and to provide professional training services to industry and commerce.

Most of the activities organised by EEA to date have been at a professional level. In particular, a structured programme of distance education modules has been developed which leads to the award of an Institution graduate diploma in engineering (professional development). The modules themselves have mostly been developed, and are delivered, under contract by universities active in distance education. The first graduates from this programme received their awards in late 1993.

Attention has also been given to the needs of engineering associates, and a diploma of engineering technology programme, similarly based on distance education modules, has been developed. This programme enables associates to upgrade their knowledge and skills to engineering technologist level.

Another subsidiary company, AE Conventions, has been established to facilitate the management of conferences and workshops, with special reference to those developed by key interest groups within the Institution, including Colleges (which are discipline based) and national committees, panels and technical societies (which cover specialised technical areas). These conferences are usually international in scope and level. In addition, the divisions and groups of the Institution often conduct specialist conferences, seminars and workshops in support of local university or industry special interests.

Overall, the Institution of Engineers, Australia, strongly supports the concept of CPD and is committed to the effective and efficient provision of a networked range of CPD opportunities and programmes tailored to the specific emerging needs of its members.

12.6.3 South Africa—CPD opportunities beckon

The economic growth and future of South Africa are directly linked to the level of technology training and the expertise available for the creation of new products. Cape Technikon, in Cape Town, is one of 15 Technikons (technical polytechnics) in South Africa providing full time national diplomas and, more recently, degrees in engineering, natural sciences and commerce.

Cape Technikon's Bureau of Continuing Education has a key role to play in addressing the backlog of education provision in South Africa. Among its wide range of CPD provision are short courses in construction management, plastics technology, packaging technology and waste management. All of these courses are offered at the request of professional bodies and institutions and often in collaboration with them.

The Technikon's policy is to accommodate students with diverse backgrounds, such as from disadvantaged communities, by means of special programmes—for example using modular courses and extended curricula. CPD courses can be tailored for employees with many years of experience but limited academic knowledge.

Research and development

Continued progress is being made with broadening of the significant community-driven applied research base of Cape Technikon. In order to stimulate technology transfer, research and educational exchange agreements have been signed with the University of Northumbria in the UK, the Swinburne University of Technology in Australia, two German Fachhochschule and one university, as well as with two Russian universities. These agreements provide for the exchange of expertise through visits by staff and students (particularly at post-graduate levels) for periods of at least three months.

Long distance education

Another feature of the pro-active approach of the Bureau of Continuing Education is the concept of lifelong learning, making use of long distance education with a difference. Television transmissions via satellite are used as a means of transmitting knowledge. The ABSA banking group is sponsoring this new concept of placing the opportunity to learn within the reach of not only large business groups, schools and public departments, but also the public. Cape Technikon's short courses can be televised and transmitted either live or by video tape.

Accrediting in-house training

The Bureau is also actively involved, in association with a private company, in accrediting in-house training courses presented by commerce and industry. The different schools of Cape Technikon play an important role, providing subject specialists to assess the academic content of these in-house courses as well as the standard of tuition. In addition, the standard of evaluation of the courses is subject to close scrutiny by Technikon's subject specialists.

12.7 How do your CPD experiences balance?

I suggest that you now give some time to thinking carefully about the lessons you might want to draw from reading this chapter. For example, you might start with analysing the current balance in your CPD activities:

- locally;
- nationally;
- internationally.

You could draw concentric circles, as in Figure 12.1, representing the amount of CPD in each area. The larger the area in the space between each circle, the greater the amount of your CPD activities in this sector.

For most of you the chance is that the area of the local circle will be much larger than the area between that and the national and international circles. Think carefully, therefore, about what you might do to increase the other two. Do not shy away from putting emphasis on the international aspect. The ease with which, for minimal cost

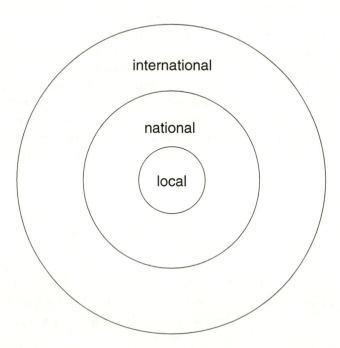

Figure 12.1   How do your local, national and international CPD experiences balance?

each month, you can access the Internet makes it extremely easy to develop your international CPD from home.

If you travel internationally for business, then why not contact the local committee of your professional institution in the countries which you visit? You could offer to give a presentation, or simply request to attend one of their meetings to extend your network of contacts.

Whatever your conclusions, make a note of them in your personal development plan and, as ever, discuss them with someone else, such as your mentor, coach or manager.

12.8 Some final thoughts

We are now coming to the end of this book, which has been an important part of my CPD and has helped me to organise many of my ideas much more effectively.

While writing this book I have been doing consultancy work in Turkey, going out there for one week in every three on a regular basis. I have been working with Gordon Williams, managing director of Gestalt Consultants, who also happens to be an IEE member and who has many years of experience at director and managing director level internationally. He is also totally committed to the concept of managers acting as coaches and has turned the performance of many companies round by applying this approach.

Our consultancy work has been with a joint venture company between one of the USA's largest companies and one of Turkey's largest conglomerates. This joint venture company was formed in 1987 and started production in 1990. Initially there were many technical problems, but the company had one very important strength—a clear vision of the value and importance of developing its people. The senior management made this belief fundamental to the way in which they solved the problems that they faced. As a result, by 1995, the productivity and quality of this joint venture was higher than in any other similar plant operated by the American partner anywhere in the world, including in the USA—and this with managers and employees who, with one exception, are Turkish. In addition, the safety record of the site is the highest in Turkey.

Despite this outstanding record, the joint venture's senior management was still not satisfied. The task for Gordon Williams and myself has been to help them to improve even further—not an easy job! And yet the learning and coaching skills in the company are so strong that one to one coaching sessions of managers and supervisors (one of our many activities there) have very quickly produced a wealth of new ideas, commitments and actions. Almost all of these ideas came from the company staff themselves, demonstrating just how powerful is the simple skill of listening. It is highly stimulating to be working with people who have so much energy, dedication to their company and trust in the management. As was described in Chapter 7, in one case it was possible, using coaching, to reduce the changeover time for two large reaction vessels from 18 days to six days, and then down to three days!

It has also been an immensely valuable part of my own CPD. I have learnt a vast amount through working with a world-class expert like Gordon, as well as by working in a world-class company such as this one in Turkey.

Yet if I were to mention the area of technology of the company and to ask you where you thought the world's most effective plant was, you would probably guess at the USA, Japan, Germany, or even conceivably the UK. It is highly unlikely that Turkey would have been on your list.

Be aware, therefore, that the world can be a surprising place in terms of your CPD. There are few places, other than under totalitarian regimes, where it is not readily possible to raise the performance and self respect of all employees, particularly in engineering.

My final tale, however, is of a nonengineer, simply because I find it so inspiring. In May 1995 I presented a paper at the 6th World Conference on Continuing Engineering Education in Brazil. After the conference my wife and I went on a sightseeing trip to the Iguassu Falls, the third largest falls after Victoria and Niagara. Our guide for the day was a Brazilian called Jim, who spoke excellent English. Jim explained that his father had been a poor farmer and that he had travelled around Brazil when young from farm to farm, wherever his father could find employment. Attending school was almost impossible, but at the age of 12 his father said to him that he was taking him to São Paulo, the major industrial centre in Brazil, for four years' proper education. After this minimal schooling, Jim obtained a production job in an American-owned factory. He noticed that many American managers visited the plant, but always needed interpreters since none of the plant managers could speak English. He therefore determined to learn English himself, but was ridiculed by his supervisors when they heard that this largely uneducated worker had such an ambitious aim. Nevertheless Jim was determined and purchased teach yourself English books and language tapes. One day he found that he was able to engage one of the American visitors in conversation in English, much to the amazement of both the American and his own managers. After this success, Jim decided that he wanted to become a guide to English speaking tourists in Iguassu and he went on to do just that—and very successfully. However, he has not just stopped there. In Iguassu he noticed that there were many German and French speaking tourists so, using the same methods that he had used to teach himself English, he went on to speak German and French; now he guides many German and French speaking tourists who all congratulate him on his fluency.

I find this inspiring simply because so many of us have had the great privilege and advantage of many years of expensive and formalised education and training. And yet, compared to this relatively poorly educated Brazilian, we release so much less of our inherent potential.

I hope, therefore, that this book has helped you to realise what you could really do to release and realise more of that enormous potential which you unquestionably have so far failed to tap.

Please contact me with your own experiences and ideas. I look forward to adding you to my own personal network of contacts and learning as much as possible from you. My e-mail numbers are: 100524.106@compuserve.com and jlorriman@iee.org.uk. Otherwise, send me a letter care of the IEE's career development department.

Good luck. Go for it! Remember that the only barriers stopping you from achieving what you really want to achieve are in your own brain!

References

1 LORRIMAN, J.A. and KENJO, T.: 'Japan's winning margins— management, training and education' (Oxford University Press, 1994)
2 LORRIMAN, J., YOUNG, R., and KALINAUCKAS, P.: 'Upside down management—revolutionizing management and development to maximise business success' (McGraw-Hill, 1995)
3 HANDY, C.: 'The age of unreason' (Arrow Books Limited, 1990)
4 HANDY, C.: 'The empty raincoat' (Arrow Books Limited, 1995)
5 HAMEL, G., and PRAHALAD, C.K.: 'Competing for the future— breakthrough strategies for seizing control of your industry and creating the markets of tomorrow' (Harvard Business School Press, 1994)
6 'A call to action—continuing education and training for engineers and technicians'. The Engineering Council, 1986
7 'Report of a pilot study—continuing education and training'. The Engineering Council, 1991
8 'Competence and commitment—the Engineering Council's proposals for a new system of engineering formation and registration'. The Engineering Council, 1995
9 'Finding the time—a survey of managers' attitudes to using and managing time'. The Institute of Management, 1995
10 HAMMER, M., and CHAMPY, J.: 'Reengineering the corporation— a manifesto for business revolution' (Nicholas Brearley Publishing, 1993)
11 RICHARDS, C.: 'Using Lotus Notes 4' (Que Corporation, 1996)
12 THOMAS, B.J.: 'The Internet for scientists and engineers' (IEE and SPIE Press, copublication, 1996)
13 'Engineering our future'. Report of the Committee of Inquiry into the Engineering Profession, HMSO, 1980
14 'Skills shortages in Europe'. Industrial Research and Development Advisory Committee of the European Commission, 1991
15 'Quality and relevance—the challenge to European education'. Industrial Research and Development Advisory Committee of the European Commission, 1991

Index